WHAT REALLY HAPPENS WHEN YOU DIE?

WHAT REALLY HAPPENS WHEN YOU DIE?

COSMOLOGY, TIME AND YOU

ANDREW MCLAUCHLIN

This edition published in 2018 by Sirius Publishing, a division of Arcturus Publishing Limited,
26/27 Bickels Yard, 151–153 Bermondsey Street,
London SE1 3HA

ISBN: 978-1-78428-456-5
DA004886UK

Printed in China

CONTENTS

Behind it all
Is surely an idea so simple,
So beautiful,
So compelling that when –
In a decade, a century
Or a millennium –
We grasp it,
We will all say to each other,
How could it have been otherwise?
How could we have been so stupid for so long?

– John Wheeler, Theoretical Physicist (1911–2008)

INTRODUCTION

'I don't believe in astrology. I'm a Sagittarian and we're sceptics.'
Arthur C. Clarke

The biggest mystery in life is surely what happens to us when we die.

You may question the basis of astrology, but your fate may still lie in the stars or, more correctly, the universe, because you are part of the universe and so your fate is inexorably linked to its future. You may think that you will be dead and gone long before the fate of the universe could have any relevance to you, but you would be wrong, for the fate of the universe actually explains what happens to you when you die.

Ask a number of people what happens when they die and their answers will probably fall into three groups. One group will say they have no idea; another will say that there is some form of afterlife in some form of heaven; and the third will say grimly that you're dead and that's the end, ashes and zilch. All these people will doubtless have one thing in common, however – they are not enamoured of the prospect of dying.

When, many years ago, my children asked me that question, I wanted to offer them a rational explanation that might allay their fears. The question had certainly haunted me when I was very young. I remember lying in bed thinking about the distant future and wondering what happened to everything – but especially, of course, to me – eventually. As I considered possibilities, one question kept coming to me: 'And then what?' I couldn't think of an answer and, indeed, found the question very disturbing. So much so that I tried not to think about it again.

So, when the subject of death came up with my own children, I asked them, 'What was it like before you were born? Was it strange? Was it horrible?' 'No,' they said, 'it was nothing.' 'That,' I told them, 'is what it must be like when you die. Nothing. Not simply nothing to worry about. An absolute nothing. Neither life nor time. So, in a way, we have all experienced death already.' They seemed quite satisfied with that, and, I think, quite content, because they never asked again and have never showed any anxiety about death.

Whatever else may be theorized about death, there can be little doubt that that simple answer represents the 'bottom line' explanation. It's an answer that makes sense and is difficult to argue with.

But more can now be said. If I were asked that question today I would be able to give a rather different answer, thanks to cosmology.

Cosmology is the science of the history and future of the universe. The big question it addresses is what happens to the universe eventually. In toiling to find the answer, some of the greatest minds in the field of cosmology, including Stephen Hawking, celebrated author of *A Brief History of Time*, have revealed profound implications, not only for the future of the universe and the nature of time, but also for the ultimate future of humanity, for you and me, because the future of humanity is hitched to the future of the universe. Their work reveals how cosmology does indeed hold the key to that most profound question of all – what happens to us when we die? But it hasn't been revealed in a blaze of glory. It has had to be teased out, although at least one cosmologist has clearly thought about it. In a 2001 television programme, *The Extreme Universe*, pondering over the significance of an infinite universe, cosmologist Paul Davies said:

> 'If the universe is infinite in spatial extent and uniform, then it is absolutely certain that there will exist other beings identical to you and me. It is one hundred percent certain that the entire inhabitants of the earth will be repeated, with duplicate Paul Davies's…' (1)

Here, Davies was considering a universe infinite in *space*. As we shall see, he has confirmed that duplicates of ourselves will also appear in the future if *time* is also infinite.

What we find is that cosmology, like religion, offers the possibility, indeed the probability, that your death is not the end of you, but, unlike religion, it offers an answer that is distinctly earthly: the history of the universe is destined to be repeated, i*n precise detail*, indefinitely, and since our lives are an integral part of that history, *we* are also destined to repeat our lives in exactly the same detail. When we die we will effectively be instantly transported forward to our birth in the 'next' cycle of the universe, simply awaking in our mother's arms just as we did when we were born in this life. The corollary of this - it may equally be regarded as the cause - is that the dimension of time is not linear, but circular.

The idea of circularity or cyclicity is not new. The 19th-century philosopher Friedrich Nietzsche proposed a similar notion, and a number of world religions also embrace the idea, usually involving some sort of reincarnation. But none of these is a scientific theory. In cosmology there have indeed been a number of attempts to put cyclicity on a scientific footing, but all these have been in terms of cycles of the universe along a linear dimension of time. Cosmology, though often criticized for its plethora of theories, is still subject to the discipline of science and the need for evidence. What is significant here is that by adhering to that discipline we are led to a legitimate, evidenced, scientific conclusion about the dimension of time and our future.

This book follows the paths of discovery marked out by the most influential cosmologists from Einstein, via Hawking, to the present day. It explains how, from its birth in a 'Big Bang', the universe is expanding, and that there are just two possibilities for its future; it will eventually collapse into a Big Crunch or it will expand forever. As we work our way through the ebb and flow of evidence and theory, we see how, whilst one of these theories emerges as the most simple and elegant explanation of our fate described above, the other, albeit more circuitously, arrives at the same conclusion. This is a persuasive indication that the conclusion is correct.

Along the way we also gather insights into some of the most important fields of science – Einstein's relativity theory, the nature of time, the mysteries of quantum theory, the thorny questions of consciousness – what makes you *you* – and also free will. We examine Stephen Hawking's controversial 'imaginary time', including the way in which it offers an intriguing idea not only about whether parallel 'multiverses' might exist but also where. In *A Brief History of Time*, Hawking said that the ultimate aim of science is to achieve 'a complete understanding of the events around us, and of our own existence' (2). Neither his book nor subsequent research claims to have accomplished that, but this book trawls through the intriguing research to expose a previously unrecognized shortcut to those two most important elements of that understanding – what happens to us and to time itself.

We see how the theory of relativity leads to those two possibilities – that the universe will either collapse into a Big Crunch under the effect of the gravitational pull of all the matter in it – a closed universe – or it will expand forever – an open universe.

In the first, 'Big Crunch', scenario it has been shown that all the 'positive' matter–energy produced in the Big Bang could have been produced out of absolutely nothing by creating an equal and opposite amount of 'negative' gravitational energy, so that throughout the life of the universe the total amount of energy is actually zero. This means that when the whole of the universe collapses into a Big Crunch, it simply disappears to nothing again, neither matter–energy nor space. We see how Stephen Hawking constructed his 'no-boundary' model of the universe from this and from 'quantum gravity', introducing the concept of 'imaginary time' as a circular dimension within which the history of the universe, from Big Bang to Big Crunch, is confined. We see how his model gives the first indications of a fundamental circularity to time but also that it can be modified by recognizing, by the Principle of Identity, that the Big Bang and Big Crunch must be considered as one and the same point in space–time because there is nothing to distinguish the 'nothing' of the Big Crunch from the 'nothing' of the Big Bang. This principle comes from the 16th-century philosopher Wilhelm Leibniz's *Identity of Indiscernibles*, which states that no two distinct things exactly resemble one another. The corollary of this is the Principle of Identity, which states;

> '...when two things, events or series of events are absolutely indistinguishable, they must be considered to be one and the same.'

If, then, the Big Bang and Big Crunch are one and the same point in space–time, the evolution of the universe must be one of eternal expansion and collapse – Big Bang to Big Crunch to Big Bang ad infinitum *around a closed loop of time*. The dimension of time must be circular. Hawking showed that Big Bangs are 'quantum states', from which a whole range of universes must emerge according to quantum theory. These are the multiverses that figure regularly in modern cosmology – and science fiction. So from the 'next' Big Bang *our* universe will emerge from amongst those others just as it did in 'this' universe, to run its course in precisely the same detail, with our life histories precisely the same, too.

So when we die we shall start our lives over again,

apparently transported directly to our birth in the 'next' universe once more.

Whilst Hawking's imaginary time proposed a fundamental circularity, it stopped short of identifying the Big Bang and Big Crunch as one and the same point so the circularity had to remain in 'imaginary time'. It could not transfer to our real-time universe. Hence Hawking's conclusion that 'we are all doomed' (3). It now seems clear that we are not.

One problem to emerge from this scenario is the passage of hundreds of billion years from the end of your life in this universe to your birth in the next. In fact, we see that since we can have no awareness of time when we are dead, those billions of years from our death in this universe to our birth in the 'next' will not, for us, exist. We awake *instantly* in our mother's arms.

The second scenario is that the universe expands forever – it's 'open'. We look at the more recent evidence from distant supernovae that supports this second possibility of an infinite, ever-expanding universe because it shows the expansion to be accelerating and will expand forever; it is open, not closed. But 'forever' means that there is time for anything that can happen to happen, by the process of 'Poincaré recurrence'. This is the process described in popular culture by which monkeys randomly typing will eventually type out a Shakespeare sonnet. If the fundamental elements of the universe behave like the letters from the typewriter, then in an infinite universe they will eventually rearrange themselves into our universe exactly the way it is now. This is shown to predict the same outcome for the universe and our lives. We look at theories as to how these recurrences might come about.

But it leaves the question of whether the 'you' in the next and future 'Poincaré recurrence' cycles will really be you. However, when we examine what makes you *you* we find that the *you* in the 'next' and 'future' cycles must indeed be *you*.

Regarding the dimension of time, if cycles of precisely identical universe and life histories occur, we see that from *our* perspective, the dimension of time will effectively be circular.

However, the latest evidence suggests that the supernovae results are actually wrong, and we are left concluding that we are in a closed, collapsing universe after all.

This is supported by the fact that when we examine the world around us we find that physical phenomena – light, heat, sound, radiation, even matter itself – all have their basis in circularity. Circularity, not linearity, rules our world. In addition, there is evidence that the ultimate theory of the future of the universe will be the *simplest possible*. The closed, collapsing universe from nothing is far and away the simplest theory.

The conclusion is that our rebirth at our death into precisely the same life seems to be unavoidable, and that we really will have eternal life in a never-ending circular dimension of time.

CHAPTER ONE

Our universe – what it's like, and how it got to where it is today

'If the universe is expanding, why can't I find a parking space?'
Woody Allen

To understand how our universe could possibly follow either of those two fundamental scenarios – expanding and then collapsing, or expanding forever – we need to understand what our universe is like, how it began and how it appears to be evolving.

The future of humanity on earth is governed, if not by its own foolish behaviour, then by the lifespan of the sun, which in a few billion years will run out of nuclear fuel and explode in a gigantic fireball, enveloping and probably evaporating the earth and many of the other planets in the solar system, leaving the rest to freeze.

Yet our solar system, huge though it seems to us, is but a tiny speck in the whole universe. How can the nature of the universe as a whole have any relevance to us, a mere speck within our solar system, in the far and distant future when we are not there? To understand this we need first to understand the make-up and origins of the universe.

Then we shall see how the future of the universe may be very relevant to us.

A glimpse at our universe as it is now

The universe is unimaginably huge and most of it is empty space. But even so there is still a lot of matter. So how big is 'unimaginably huge' and how much matter is 'a lot'?

The universe is so immense that distances are generally measured in light years – the distance light travels in a year. Light, of course, is so fast that it makes the earth hurtling around the sun at 70,000 mph appear positively sluggish. Light travels at 186,000 miles per second, around a billion miles per hour, making a light year a distance of 6,000 billion miles. The size of the universe is measured in terms of the furthest objects we can detect. The furthest detectable stars are some 10 billion light years away, meaning a distance so great that light from those stars takes 10 billion (10,000,000,000) years to reach us. How far in miles? That's a distance of 60 billion trillion (60,000,000,000,000,000,000,000) miles. If you can't imagine that distance, don't worry, no one can. These huge distances offer us a form of time travel, for they allow us to look back into the past. Since light from a star 10,000 light years away takes 10,000 years to reach us, the images we see must be of the star as it was 10,000 years ago. This means, of course, that we cannot know what is happening to that star or any other star *right now*. Every event we observe is history, since the light we use to detect them takes a finite time to reach us.

All the matter we can see outside our solar system – though not all the matter there is – is in the form of stars clustered into galaxies. Most of the stars we can see on a clear night belong to 'our' galaxy, the Milky Way. Our sun is just one of them. The next nearest galaxy to ours is over 100,000 light years away. If it takes light 100,000 years to get there, then there is no chance of you and me getting there, and there is really little likelihood of any human contact for the simple reason that our species is likely to be long gone by the time any near-light-speed transport could get there. Inter-galactic space travel is, and is likely to remain, science fiction. Travel within our own galaxy is a little more conceivable, though our nearest stellar neighbour, the triple star system Alpha Centauri, is still 4.3 light years away – a distance of 25 trillion (25,000,000,000,000) miles. Not exactly walking distance but maybe possible if near-light-speed transport becomes a reality. Communication would obviously be more practical than travel, but even this would prove problematic. Since radio waves travel at the same speed as light, messages would take 4.3 years to wing their way to Alpha Centauri. And the response would take 4.3 years to come back. Since answers to any question would be 8.6 years out of date, conversation would be difficult.

And how much matter is there in the universe? Our Milky Way is a disk-shaped cluster of stars some 100,000 light years across. Our sun and its planets lie towards the edge. All the stars we see on that occasional clear night are our very 'near' neighbours in the Milky Way. The rest are so far distant that they appear merely as a milky coloured pathway – hence the name. Yet they are still part of our galaxy. Altogether there are over 100 billion stars in the Milky Way and our sun is just one of them.

Until the 1920s the Milky Way was thought to contain all the stars in the universe. However, we found that our galaxy, with its 100 billion stars, is but one of 100 billion other galaxies evenly spread across the universe. This puts the number of stars in the universe at some 10 billion trillion, a number far too big to comprehend. We could say that it is more than the number of grains of sand in all the deserts and beaches in the world – a calculation provided by Australian astronomer Simon Driver (1) – but that might not help us to appreciate it.

And where did all this matter come from? Until early last century, it was believed that the universe, as we see it now, had existed forever, or at least since the Creation. The evidence now leads to two theories: the first is that the universe blew up out of a microscopic 'cosmic seed'; that some 13.7 billion years ago everything that is now in the universe was contained in a space so small that it would pass through the eye of a needle as easily as a fly passes under the San Francisco Bridge. The second, and for good reason the most likely, is that everything in the universe actually came from nothing at all.

At this point, the more rational of us start to raise our eyebrows. We might accept the ludicrously large number of stars even though we haven't seen them. We might accept some kind of primeval explosion billions of years ago even though we weren't around then, but the idea that all those stars, each much like our sun, could somehow have been contained in a microscopic volume or, more absurdly, have actually come from nothing at all? This is a good example of our common sense letting us down when we are looking at extremes. And as for it all coming from nothing at all – well Stephen Hawking, for one, believes so.

The early view of the universe

Until the start of the last century, on the basis of biblical evidence, the universe was widely believed to be 6,000 years old. We might now scoff at such an absurdly small figure, but we must bear in mind that in the absence of both scientific enquiry and the access to it that we now take for granted, our knowledge of the world was based solely on what we could see, together with 'received wisdom', which in those days was essentially biblical. The birth of the universe was therefore seen as a religious rather than a scientific event. Only when geological evidence showed that the earth was millions, not thousands, of years old was this biblical age rejected by most scientists. Nevertheless, until well into the last century, most people, scientists included, viewed the universe as unchanging; it had been created much as it is now and would remain so forever. Even though the stars were recognized as hot bodies that might eventually cool down, the universe as a whole was viewed as eternal and unchanging. After all, we can see no change in the pattern of stars from year to year, generation to generation. What other conclusion could we draw?

How Einstein's Theory of Relativity launched a new era in cosmology

In 1905, the German-born physicist Albert Einstein presented his special theory of relativity. Einstein is usually shown as a wise old man with grey hair – the archetypal image of a genius. In fact, when he presented his theory, he was a young man, brisk and dapper. He was, however, certainly a genius. His ability to view the commonplace from a completely different perspective and to follow the implications through from mathematical complexity to elegant physical simplicity was extraordinary. He showed that space and time were not the separate, unconnected concepts that common sense tells us they are, but different dimensions of a single quantity – we now call it space–time. Einstein showed how space and time are related. The beginning – or end – of space is the beginning – or end – of time. This is central to our ideas about the origin of the universe, together with Einstein's other remarkable discovery that appeared in his special theory, namely that matter and energy are different forms of the same thing, a relationship expressed by his famous equation $E = mc^2$.

These scientific discoveries seemed to fly in the face of common sense. How could time, something we understand intimately, be related to space, which we know even better, without our being aware of it? Even more implausible, surely, is the suggestion that matter is a form of energy. We know that we can get energy out of matter through chemical reactions, but Einstein showed that matter actually *is* energy! This was the first of many adjustments that science obliged us to make to our sense of reality.

Twelve years later, Einstein presented his general theory of relativity. The special theory of relativity was so-called because it considered a special case in which the effects of gravity were ignored. In his general theory, Einstein introduced gravity into his equations and showed that not only were space and time aspects of one entity – space–time – but also that space–time was actually shaped by gravity and not fixed and permanent, as had been assumed. Gravity, said Einstein, is not an ordinary force, but the manifestation of space and time being bent around matter. He based his theory on the fact that it would be impossible for anyone to distinguish between the effect of gravity on earth and the effect of being accelerated at 'g' (10 m/33 ft per second per second) in gravity-free outer space, or between free-falling on earth and floating in outer space. In this, Einstein was also employing a principle that is central to this book, the Principle of Identity. When it is impossible to distinguish between two things, events or series of events they must be one and the same.

What the general theory also showed was that the universe could not be static. It must be expanding or contracting. But so strongly held was the view that the universe was unchanging that Einstein couldn't accept his own findings and actually adjusted them, ensuring that he became famous not only for the most important developments in physics, but also for claiming to have made 'the biggest blunder of my life' (though there is doubt about the provenance of this quotation). This is a reminder that even the greatest minds can make big mistakes. Though Einstein closed his mind to this one aspect of his findings, others didn't. A Russian physicist, Alexander Friedmann, showed that Einstein's theory led to two models of an expanding universe. In one, the galaxies start out with zero separation, expanding to a maximum size, and then contracting again to the original state of zero separation – a 'closed' universe (Fig 1.1). In the other, the galaxies continued to separate forever – an 'open' universe (Fig 1.2). Later, a variation was added by French cosmologist and priest Georges Lemaître, of a universe that would be just open but remain on the point of collapse. This was called a 'flat' universe.

All these models, however, quite clearly showed that the universe was currently expanding. More importantly, and remarkably, relativity showed that the expansion was not a matter of the galaxies moving apart through space but of space itself expanding and carrying the galaxies along with it, like spots on an inflating balloon. Equally noteworthy was another 'relativity model' created by Dutch astronomer Willem de Sitter. In this model of a totally empty universe, it was found that when particles of matter were mathematically 'sprinkled' into the equations they rushed apart. And in a perfect example of scientific methodology, de Sitter also predicted that the expansion should be detectable in the real universe. The theory could be tested. But how? In fact it was surprisingly easy. Light, like sound and every other physical phenomenon, has wave properties and, like sound, is characterized by its frequency. But whereas we distinguish different frequencies of sound by their pitch, we distinguish different frequencies of light by their colour. More importantly, by using an instrument called a spectrometer, the different colours – frequencies – of light can be displayed as a spectrum with each colour occupying a specific position relating to its frequency.

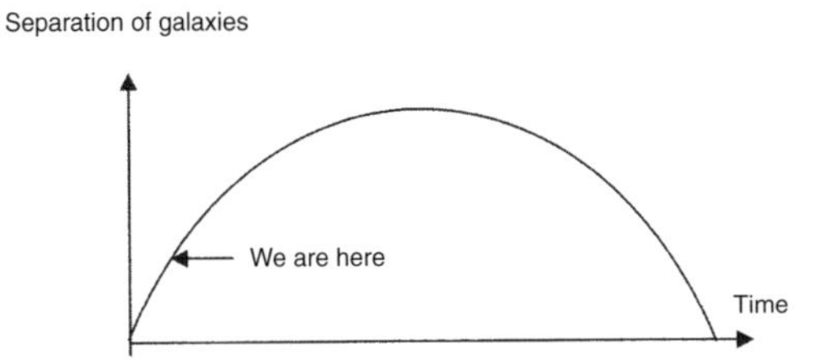

Fig 1.1 A 'closed' universe – one that collapses back into a 'Big Crunch'

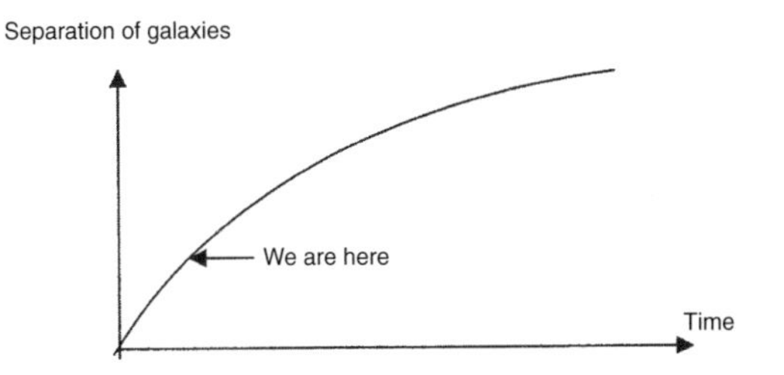

Fig 1.2 An 'open' universe – one that expands forever

Raindrops act as mini-spectrometers to produce a rainbow if the sun is behind them. Just as the sound waves from a racing car are stretched as it speeds away from us, making the sound lower in pitch – a phenomenon known as the Doppler Effect – so the frequency of light from a source moving away from an observer is stretched, shifting it towards the longer wavelengths at the red end of the spectrum. De Sitter predicted that the frequency of light from stars would be changed in this way if they were receding in an expanding universe. Not only that, said de Sitter, but the effect would be proportional to their distance from us.

In 1929, Edwin Hubble, the astronomer who is commemorated by the space telescope that nearly a century later sends us images of distant parts of the universe, made observations of stars called Cepheids that confirmed this red shift. In particular, he showed that the further away the star appeared to be, the more it was red-shifted – in other words the stars were accelerating away from us. And this acceleration was uniform, wherever he looked in space. In doing so, Hubble not only confirmed de Sitter's theory but also laid the foundations of modern cosmology. His detection of the change in the frequency of light as predicted, signalled general acceptance that the universe is expanding.

How it all started – the Big Bang

The discovery that the universe was expanding, even with the theoretical framework offered by relativity, did not, however, fully explain the processes involved. It was accepted that the universe had started out smaller and denser than it now is, but how small? How dense? And how exactly did it all start?

In 1948, Russian-born American physicist George Gamow presented a coherent and carefully crafted model of the universe that gave an exciting and detailed answer to those questions (2). The universe, he showed, was born out of a minute but incredibly dense fireball of radiation and subatomic particles that blew apart in a cataclysmic explosion. As it did so it cooled and condensed into clouds of the simplest forms of gaseous matter, which in turn eventually collapsed under their own gravity to become the galaxies of stars we see today. Gamow's model became known as the 'Big Bang' model of the universe.

There is some debate as to whether it was Gamow who first coined the term 'Big Bang' or Fred Hoyle who, challenging Gamow's theory, used the term in a derogatory way. Hoyle himself championed an alternative 'Steady State' theory that discounted the Big Bang and suggested that the universe was much the same now as it was in the infinite past and will continue that way into an infinite future, with new matter being continuously produced in the spaces between the expanding galaxies. Subsequent research, discussed later, cast serious doubt on the Steady State theory and it has been largely consigned to history. Hoyle maintained that he did not use the term 'Big Bang' disparagingly but to make the concept clear to the public – which it did. Whilst Gamow may not have been a larger-than-life character like Stephen Hawking or Hoyle (who, incidentally, wrote a large collection of quality science fiction), he had a sense of fun. His Big Bang research was carried out with a student, Ralph Alpher. Gamow persuaded a prominent nuclear physicist, Hans Bethe, to let him use his name in the published research paper, so that the authors appeared as 'Alpher, Bethe and Gamow' as in the first three letters of the Greek alphabet, alpha, beta and gamma, reflecting the theme of their paper – the beginning of things.

The formation of matter

The central strength of Gamow's model was the way it explained how the universe came to be made up of hydrogen and helium, the two simplest forms of matter, in the proportions we see today, roughly 75 per cent hydrogen and 25 per cent helium. Equally significant, however, was

Gamow's prediction that if there really was a Big Bang, there should be a residue, an echo, still detectable in space in the form of microwave radiation. In 1965, two American researchers, Arno Penzias and Robert Wilson, whilst trying to detect weak radio signals from space, were irritated to find their sensitive equipment picking up extraneous background noise. In desperation they even cleaned pigeon droppings from their big horn antenna. They did not realize that they had detected the cosmic background radiation that Gamow predicted until they contacted American physicist Robert Dicke, whose team had been looking for the very same thing. Dicke then confirmed this discovery. Gamow's Big Bang theory was home and dry. In 1992, the COBE (Cosmic microwave background explorer) satellite, the brainchild of cosmologist George Smoot, amazed the scientific world by probing deeper into space, and therefore further back in time, than had previously been possible. In doing so it was actually able to measure the minute temperature variations in the microwave radiation, thereby not only confirming Penzias, Wilson and Dicke's findings, but also providing further information about the earliest stages of the formation of the universe.

With the knowledge of the formation of stars, developed over the previous decade, it was possible to estimate when the Big Bang occurred. In the 1930s the nature of nuclear reactions was becoming understood and it was shown that stars were not simply cooling sparks of white hot matter, but were gigantic nuclear reactors fuelled by the nuclear energy released as gravity pulled the stellar material together. The processes involved allowed a calculation of the age of the universe. This age, originally set at 10 billion years, has been steadily revised upwards as new information has come to light and is now (2015) generally thought to be 13.7 billion years, though some scientists think it may be even older than that – maybe 15 or even 20 billion years.

By the end of the 1960s, then, there was general acceptance of a very detailed Big Bang model in which a 'cosmic seed' containing all the matter and energy in the universe compacted into the smallest volume exploded in a fireball at a temperature of many billions of degrees. The fireball was initially a 'soup' of radiation and subatomic particles of enormous density. Within minutes the universe expanded and cooled to a modest few million degrees, allowing the soup to condense into the simplest and most common forms of matter we see in the universe today, hydrogen and helium. Over the next billion years or so the clouds of gas coalesced under the effect of their own gravity to form heavier matter, and ultimately to matter so compacted that nuclear reactions occurred that enabled them to burn white hot. These are the stars and galaxies that make up the universe as we now know it.

The inexorable pull of gravity between the exploding matter–energy that began at the moment of the Big Bang, would, of course, slow down the expansion, posing the question as to whether the expansion will eventually stop and then go into reverse, contracting in the manner predicted by the first of Friedmann's theoretical 'relativity' models mentioned earlier. Until 1998, many cosmologists, including Stephen Hawking, believed it would. This contraction would then end in the opposite of the Big Bang – a total collapse into a 'Big Crunch', signalling the end of the universe.

Relativity appears again and predicts a 'singularity' at the Big Bang

In the early 1960s, there was a surge of interest in astrophysics with the discovery of quasars, star-like bodies thought to have been formed in the Big Bang. When it was shown that they could have been produced by gravitational collapse, physicists dusted off their copies of Einstein's general theory and began to apply it to cosmology once more.

In 1965, mathematician Roger Penrose applied relativity theory to a star collapsing under its own gravity when it ran out of nuclear fuel. He showed that if it collapsed beyond a certain point it couldn't re-expand. Instead, it would collapse into a 'singularity', a point of zero size and infinite density at which the laws of physics break down because the mathematics cannot deal with infinite quantities (3). Penrose was in fact describing the formation of what were, four years later, to be called 'black holes' by cosmologist John Wheeler. A young Cambridge PhD student, desperately searching for research material for his thesis, came across Penrose's work and wondered if it could be applied to the universe as a whole. His resulting work ensured that he collected his PhD in fine style and determined the direction of his future research. The student was Stephen Hawking, and he went on to show, with Penrose, that if General Relativity is correct, the universe must have begun with a singularity, a point of zero size and infinite density marking the beginning, not only of space, but time. Everything we now know to be in the universe was compressed into this singularity.

This reveals another strange feature of the origin of the universe. The Big Bang is generally understood to be a primeval explosion of matter within some sort of region of empty space. Relativity theory and Hawking and Penrose's work, in particular, show that this is not so. The Big Bang was not only the origin of matter and energy, but also of space itself – and time. It was in fact the origin of everything.

Ironically, within a few years, Stephen Hawking abandoned the 'singularity' theory because it caused the laws of physics to break down. If they break down there, he thought, they could break down anywhere.

CHAPTER TWO

The origin of the universe – a universe from nothing?

'The surest sign that intelligent life exists elsewhere in the universe is that it has never tried to contact us.' ***Bill Watterson***

The view that the universe began with a Big Bang explosion from which all the stuff we now see in the universe blew up from a 'singularity' of concentrated matter–energy became the accepted picture until the late 1960s, when the first suggestion was made that the universe could have appeared from nothing at all.

The idea can be traced back to a conversation between George Gamow and Einstein during World War II when Gamow was working in the USA on the Manhattan atomic bomb project – and Einstein wasn't because of his nationality. According to Gamow (1), he mentioned to Einstein that a colleague, Pascual Jordan, had shown how a star could be created out of absolutely nothing because its positive matter–energy could be exactly cancelled out by the negative gravitational energy of this matter. Einstein was so stunned by the idea, Gamow says, that he stopped in the middle of the road and was nearly run down. This episode was to expose a largely unrecognized property of gravitational energy.

Einstein's theory of relativity may not be widely understood, but his equation, $E = mc^2$, has almost become part of everyday language. Its meaning, however, is certainly not part of everyday understanding. It says that matter and energy are equivalent; they are actually different forms of the same thing. To understand this we need to look at the nature of energy.

Energy and how it behaves

Energy is the ability to do work. There are different forms of energy – heat, electrical energy, chemical, mechanical and nuclear energy are the most common. What they all have in common is the ability to do work. A gas flame can turn water into steam whose pressure can turn a turbine. Electrical energy can turn an electric motor and so on. But one form of energy is special and that's gravitational energy. When you lift a weight, the energy you use is given to the weight and it can be reclaimed by making the weight do some work as it falls. An old-fashioned clock powered by weights works this way. You lift the weights, which then fall, their gravitational energy being transformed into mechanical energy by the gear wheels in the mechanism to move the hands. Gravity is not, however, merely the pull of the earth on all the objects near it. Every lump of matter in the universe exerts gravitational attraction towards every other lump. So whilst the earth is pulling you, you are pulling the earth. Not only that, the people standing next to you are pulling you towards them. Unlike the pull from the earth, however, the gravitational attraction between people is so small as to be unnoticeable because gravitational attraction depends on the mass of the objects involved and is only noticeable with objects as big as mountains. Even the pull of an object as large as the moon is only really noticeable to us when it pulls on the sea and, with a little help from the sun, produces tides. The sun, as the most massive object in the near universe, dominates our region of space, its gravitational pull on the planets keeping them in orbit and holding the earth in the perfect position for the support of life.

It is a characteristic of every form of energy that it can change from one form into another. Heat energy can be turned into mechanical energy in a car engine, chemical energy can be turned into heat in a fire or electrical energy in a battery and so on. And every event in nature, from the smallest twitch of an atom to the outpouring of energy from the sun, happens as a result of energy changes. The history of the universe is the history of energy changes. It is the sequence of energy changes that represents the dimension of time.

The most common events involve changes into heat. All forms of energy convert readily, and in fact inevitably, into heat. Sometimes the problem is how to stop this happening. However, we can never end up with more energy than we started with. This is a fundamental law of nature, the law of the conservation of energy. It says that energy cannot be created or destroyed, only changed from one kind to another.

In his ground-breaking special theory of relativity, Einstein showed the most remarkable energy change of all – matter into energy. He showed that matter is actually another form of energy. We can measure it in mass units such as kilograms or in energy units such as joules or calories. The equation $E=mc^2$ converts one to the other. It tells us how much energy 'E' a mass 'm' is equivalent to. The equation includes a constant 'c', which is the speed of light. This number is so big – and 'c squared' (c^2) even bigger – that a small amount of matter is equivalent to a huge amount of energy. The most spectacular confirmation of this is seen in nuclear reactions in which tiny amounts of nuclear mass are converted into enormous amounts of heat energy.

Just as matter can be converted into energy, energy can be converted into matter. George Gamow, in his Big Bang theory, described how the matter that ended up as stars began as a 'soup' of radiation energy and sub-atomic particles that condensed into pure matter as the universe expanded. In fact, the particles under those earliest extreme conditions represented only a small part of the total energy involved. The story of the birth of the universe is very much one of a conversion of energy – radiation energy – into matter–energy. $E=mc^2$ tells us that an awful lot of radiation energy was needed to create all the matter we know to exist in our universe. But this energy, like all other forms of energy, had to come from somewhere in the first place. It was assumed that it was all present at the very beginning, as a 'singularity' of zero size and infinite density as Penrose and Hawking suggested. But there is another possibility which, like relativity theory, comes from the nature of gravity and has had a major influence on the progress of cosmology: namely that the universe came from nothing at all.

Positive matter energy and negative gravitational energy

Gravitational attraction – the pull of every object towards every other object – is acting on all the matter–energy blown apart in the Big Bang in trying to pull it back together again. All this matter–energy, therefore, has gravitational energy, the sort of energy stored in a car at the top of a hill that can cause the car to travel downhill without the use of its engine. Gravitational energy is special. If we call the energy of matter 'positive energy', then gravitational energy can be shown to be negative; that is, it is has the ability to cancel out positive matter–energy. This is yet another strange but solid scientific fact and it is of fundamental importance here. The explanation is straightforward. Gravitational energy results from work done against the force of gravity. Imagine two bodies close together floating in space. Gravitational attraction is pulling them together. To pull them apart requires work against gravity. They thus gain gravitational energy. Now the pull of gravity between two bodies reduces as their separation gets larger. And it reduces quickly because gravity obeys an 'inverse square' law that says that doubling the distance reduces the force by a quarter. At ten times the distance, it is reduced a hundredfold – so gravitational attraction falls away very quickly with separation. This means that if two objects are so far apart that the force cannot be felt (that is at infinite separation), the gravitational attraction must be zero, and so their gravitational energy must be zero. But then, as they come together, their gravitational energy must reduce, just like the gravitational energy of a car rolling down the hill. But with zero energy to begin with, losing energy means that the gravitational energy must become negative. At any distance less than infinity gravitational energy must be negative. And it must become more negative with decreasing separation. For the universe this means that negative gravitational energy offers a way in which it could have begun from nothing, rather than the infinite compressed energy proposed by Penrose and Hawking.

It came about this way: before his walk with George Gamow, Einstein had been well aware that quantum theory – the theory of sub-atomic physics – allowed individual atomic particles to be created from nothing by 'borrowing' energy from the vacuum around them for a very short time. 'Empty' space is not, in fact, truly empty at all. Quantum theory shows that it is in fact a seething mass of subatomic particles bursting in and out of existence, living not only on borrowed time but on borrowed energy. The borrowed energy is paid back quickly and the particles disappear again. The conservation law of energy allows this to happen so long as the energy is paid back quickly. The more energy that is borrowed, the quicker it must be paid back and therefore the shorter the life of the particles produced. The idea that Gamow passed on to Einstein, however, was different. The suggestion was that matter could be created from nothing on a cosmological scale, for the lifetime of a star.

The idea that so stunned Einstein was that the positive matter–energy of a whole star could be produced from nothing at all if, at the same time, an equal amount of negative gravitational energy could be produced in the matter itself. It was only a matter of time before someone suggested that the whole universe could have been created this way.

It appears that the first person to suggest the notion of the whole universe being created from nothing was a young American assistant professor, Edward Tryon, at a lecture given by cosmologist Dennis Sciama in the late 1960s. During a pause, Tryon blurted out 'maybe the universe is a vacuum fluctuation'. (That is, an extreme case of the same process that allows atomic particles to appear from nothing.) His comment was evidently greeted by laughter, so much so that the researcher was acutely embarrassed. However, the theory clearly evolved in his mind and, eventually, he published the fully worked-out idea in a scientific paper in 1973 (1).

He proposed that all the matter–energy in the universe could have been created from nothing, as an exceptional quantum fluctuation but with the energy 'borrowed', as it were, by the process described to Einstein by Gamow, from gravity, with no need to suspend the conservation law. The process would be that, as the matter–energy formed at the Big Bang is blown apart, it acquires gravitational energy, the energy of its own gravitational attraction trying to pull it all back together again. The positive matter–energy is created along with or, rather, 'out of', an equal and opposite quantity of negative gravitational energy. At all times these balance, so that the total energy remains at zero. In this way the universe is created instantly. It's rather like wealth, which can be created instantly by borrowing money and going into debt. Although when we borrow money we have it to do what we like with, we always carry a debt equal to the amount we borrowed, so that even if we spent none, our net audited wealth would be zero. In the case of matter creation, the debt is gravitational energy. If we were to do an 'energy audit' on a universe created in this way, we would find that at any time in its history its total energy would be zero – the positive energy of the matter in the universe would be exactly cancelled out by its own negative gravitational energy. All the matter–energy created at the Big Bang could have been created, like wealth, from nothing at all simply by creating an equal and opposite amount of gravitational energy 'debt'.

In this 'universe-from-nothing' model, the Big Bang would therefore simply be the means by which the gravitational energy is produced. The matter created must blow apart to create the gravitational energy it was made from. One of the biggest questions in cosmology has been why the Big Bang occurred in the first place. If, as Gamow thought at the time, and Penrose, Hawking and most cosmologists have thought since, all the matter in the universe began life concentrated in an infinitely small space, bound together by an unimaginably immense force of gravitational attraction, why would it suddenly blow apart? One theory has been to propose the antithesis of Einstein's 'biggest mistake', the cosmological constant he introduced into his equations of relativity to stop them showing the universe expanding. The theory is that a new 'cosmological constant' would make high concentrations of matter unstable and blow apart in big bangs. But in a universe from nothing there is no need to explain why the original matter blew apart, because there was nothing there to blow apart. The Big Bang was, in fact, simply the necessary process of producing the gravitational energy from which the matter–energy was created.

A universe from nothing is not infinite

Such a 'universe from nothing', however, would produce a universe that is closed, that is, one in which all the matter is eventually pulled together again by its own gravity so that it eventually collapses into a 'Big Crunch' – Friedmann's first model from relativity theory. In 1973, when Edward Tryon proposed his universe from nothing, the prevalent view was that the universe was not going to collapse but would expand forever; it was an 'open', not 'closed' universe. It was also difficult to understand why a mega-fluctuation big enough to create a whole universe should occur when the process normally involves sub-atomic matter. On top of this, Tryon's theory did not address the two problems that had recently emerged from the Big Bang theory. The first has become known as the 'horizon' problem. From a cataclysmic explosion there should be much more irregularity in the universe than there seems to be. In an explosion the bits fly away in a highly irregular fashion, so different parts of the background radiation should be at different temperatures. Yet the extremes of the universe actually look very similar in all directions. Apart from the minute variations that confirm the Big Bang theory, the temperature of every part of the universe is virtually the same. Temperature equalization can only occur if there is some means of transferring heat from warmer to cooler areas, but a rapid expansion of the universe, such as Tryon proposed, would have prevented any such heat travel. A viable 'universe from nothing' theory must therefore involve all parts of the universe being in contact long enough for temperature equalization to take place.

The second problem was Lemaître's variation: that the universe appeared to be in this most unlikely state of 'flatness', on the finest of dividing lines between collapsing under the influence of gravity and expanding forever. Why did the universe not quickly collapse under its own gravity soon after the Big Bang? A slick solution is that if the universe were other than flat, we wouldn't now be here to be asking questions about it. The rate of expansion has thus been such as to create the conditions conducive to life that can evolve sufficiently to ask questions about the origin of the universe. This is the 'anthropic principle' – 'it's like it is because if it weren't we wouldn't be here to see it'. It isn't wrong, but it isn't very satisfying either. To scientists who want theories that network other theories and have predictive powers, the anthropic principle is a bit of a cop-out. As a result, Tryon's theory was a little before its time and did not achieve the recognition it deserved.

Inflation theory – the ultimate free lunch

A few years later, when the possibility of a closed, collapsing universe became more widely considered, a more elegant solution appeared that solved the two problems and pushed Tryon's theory into obscurity. It was offered by American physicist Alan Guth in 1980, who proposed a process he called 'cosmic inflation'. In this theory, for a brief instant before expansion all the regions of the new universe were in contact to equalize their temperature. That split second was sufficient to ensure that all the bits that flew apart were together long enough to settle to the same temperature before they went their separate ways. After this point the size of the universe multiplied exponentially – by a factor of 1,050 in 10–35 sec – like the chain reaction in a nuclear bomb, only inconceivably faster. And the rapid expansion was sufficient to 'flatten' the universe. Guth likened the process he discovered to blowing up a wrinkled balloon. No matter how wrinkled and irregular it starts out, blowing it up irons out the wrinkles and, to a tiny insect on the surface, the balloon very quickly looks not only smooth but quite flat (2). Significantly, Guth's model, like Tryon's, was also proposing a universe created from nothing, but with a more solid theoretical foundation. Guth called it 'the ultimate free lunch'. This time the cosmological establishment was more receptive.

A universe from nothing also does away with the problematic 'singularity' predicted by Hawking and Penrose, the point that so bothered Hawking because it meant that the density of the universe is infinite, its size is zero and the laws of physics must break down. In fact, a universe from nothing explains how relativity theory got it wrong in the first place. It assumed that there was always some matter to begin with. Therefore, if its size was zero its density must be infinite, and the point must be a singularity. The only way that the density of matter of zero size could be anything but infinite is if there was zero matter as well.

The origin of 'nothing'

One problem with both Tryon and Guth's theories was that the 'nothing' from which the universe was made was a vacuum, the stuff left in space when all the matter and energy have been taken out. Guth showed how the universe could have emerged from a small seed of vacuum over a billion times smaller than the nucleus of a hydrogen atom. But the mere absence of matter or energy does not make what's left 'nothing'. Empty space is very much something. It is a 'quantum foam' of activity, as John Wheeler called it, with sub-atomic particles continually materializing and disappearing. Relativity showed Hawking and Penrose that the universe began with matter–energy and space compressed into a point of zero size. If there was no matter–energy then there was still the question of space. Even if the universe began from the most minuscule speck of vacuum, the origin of this vacuum requires explanation.

In 1982, Russian physicist Alexander Vilenkin offered a solution. He picked up Tryon's theory and incorporated it into Guth's notion of inflation, proposing that the universe was created out of absolute nothing – not only no matter or energy, but no space, time or vacuum either (3).

Vilenkin's theory was even simpler and more elegant than Guth's: that the whole universe, matter, energy, space and time, all came from absolutely nothing at all – no matter, no space and no vacuum. It made several proposals: that the positive matter–energy of the entire universe is indeed created from an absolute nothing along with a negative 'debt' of gravitational energy; that it lives its life in this debt – its net energy always zero, and that when the negative gravitational energy becomes equal to the positive matter–energy, at maximum expansion, it finally collapses under gravity, with matter, energy, space and time disappearing once more into nothing in the Big Crunch. At that point the 'loan' is cleared and the 'debt' paid off, leaving no sign that a universe ever existed. Vilenken's closed-universe model provides an attractive explanation of the origin of the universe. In doing so it also provides an explanation of its fate, for a universe from nothing must collapse back to nothing once more in a Big Crunch. The origin of a closed universe dictates its fate just as the pull of a catapult determines the destination of the missile.

With this model we have an explanation of where all the matter from the Big Bang came from. It came from absolute nothing, no matter, no energy, but above all, no space. There was no need to explain why the Big Bang began expanding. Expansion was the only way that the necessary negative gravitational energy could be produced. There was no need to explain why the universe began at such a phenomenal temperature. It was at such a temperature, not because the universe was hot, but because the universe *was heat*. The initial universe was not something that was incredibly hot. The positive energy produced was all in the form of radiation. Since it occupied an incredibly small volume initially, the universe was simply an incredibly high concentration of radiation energy. Such a high concentration of radiation means a lot of heat. As the universe expanded, the temperature didn't fall by the universe losing heat, because, as with hot water in a vacuum flask, there was nowhere for heat to be lost to. The temperature fell because the radiation was being spread over an ever-increasing volume. The energy that was the initial universe was simply thinning out. True cooling began as the radiation 'condensed' into matter – heat radiation being turned into matter meant there was a lot less heat radiation in the universe – an awful lot of radiation energy is needed to make a little matter ($E = mc^2$ again).

And, of course there, is no need to speculate about the future of a universe from nothing. It must eventually collapse again under its own gravity into nothing in a Big Crunch.

Hawking describes the 'universe from nothing' in terms of borrowing matter–energy from gravitational energy. The concept of 'borrowing' may seem odd but it actually provides a helpful analogy, particularly for those of us who find it hard to accept the idea of creating the universe from nothing at all. In this model the matter–energy of the universe was created just as money can be 'created'. If I have no money at all, I can get some by going to the bank and borrowing it. I walk in with no money and walk out with money. I haven't stolen it. It is mine and is as real as any other money I have ever had. My money has been created from nothing by creating an equal and 'opposite' amount of debt. There is no limit to the amount of money I can obtain so long as I accept an equal amount of debt (and can convince my lender that I will be able to repay it). So it is with the creation of matter in the universe. To quote Stephen Hawking:

> 'During the inflationary period the universe borrowed heavily from its gravitational energy to finance the creation of more matter ... The debt of gravitational energy will not have to be repaid until the end of the universe.' (4)

In this monetary analogy, the amount of debt must always equal the amount borrowed. This, of course, corresponds to the law of conservation of energy. The 'debt' of the negative gravitational energy of the matter–energy blown apart in the Big Bang must be exactly equal to the 'borrowed' positive matter–energy. This means that regardless of how much matter was created, the universe will simply expand until the negative gravitational energy equals the positive matter–energy created.

However, when the point is reached at which the negative gravitational energy is equal to the positive matter–energy, the universe must stop expanding. To continue is to contravene the conservation law. And from that point the only way is back. The universe must then collapse.

'Bat and ball' inflation

One way to visualize the inflation period that kicks off the expansion is to imagine a ball being hit vertically into the air. The ball is given kinetic energy during the short period it is in contact with the bat. This corresponds to the period of inflation during which all the matter–energy was created. The amount of energy given to the ball by the impact determines how high the ball will rise because that will be exactly equal to the subsequent gravitational potential energy of the ball. The ball then rises by exactly the amount that corresponds to the energy given at impact – and no more.

And so it is with the universe, if it started out from nothing. As the matter–energy is created during inflation, and the universe expands rapidly, but not all the corresponding negative gravitational energy is produced during the inflation period. If it were, the universe would promptly collapse at the end of the inflation period. In fact, just as the ball must continue upwards at the end of its contact period with the bat until its kinetic energy is spent, so the universe must go on expanding after the inflation period – until the negative gravitational energy is equal to the positive matter–energy created. Just as the ball has 'potential' gravitational energy at the end of the period of contact with the bat, so the universe has 'potential' gravitational energy at the end of the inflation period. There can be no question of there being so little matter that the universe will continue expanding forever, for then the gravitational energy could exceed the matter–energy produced. Likewise, there can be no question of there being more matter–energy created than gravitational energy. Here, of course, the bat and ball analogy breaks down because we can actually give a ball more energy than the gravitational energy it can theoretically gain by going up forever. The ball would then reach its escape velocity and it would never come down. Since it began from nothing, the universe had no such limitless sources of energy. All the matter–energy has to be obtained from gravitational energy. Unlike the ball, the universe *must* come down.

Regardless of the amount of matter–energy created then, or whether it is dark or light, if the universe came from nothing the conservation law dictates that the universe will simply expand until an equal and opposite amount of gravitational energy is created. Then, like the ball at the top of its trajectory, it must stop and collapse back again, into a Big Crunch. The Big Crunch is inevitable in the model of a universe created from nothing.

And then? If it collapses into a Big Crunch again, it must lose all its negative gravitational energy. But the only way it can lose gravitational energy and maintain the law of conservation is for the matter–energy to disappear again. We are left with a scenario in which at the 'Big Crunch' the gravitational energy disappears and takes all the matter–energy with it. The universe disappears as if it never existed. It is payback time. In the Big Crunch, the energy borrowed from gravitational energy to create the matter–energy of the universe is paid back – in full.

This represents the simplest theory of the universe. Everything that has happened, and will happen to, and within, the universe would all be contained in that cycle of expansion and collapse. It does not tell us the specific details of every phenomenon – physics still has much to discover – but tells us the ultimate consequence of the laws behind such phenomena. It represents a 'cosmological short-cut'.

The Big Crunch will not simply be the end of matter–energy. It will signal the end of everything – matter, energy, space and time as we know it. As Paul Davies has written:

> 'A universe that came from nothing in the Big Bang will disappear into nothing at the Big Crunch, its glorious few zillion years of existence not even a memory.' (5)

To many, certainly to some seasoned scientists like Paul Davies, the prospect of a Big Crunch into oblivion is much more depressing than the big freeze of an open universe. An inexorable collapse into the fiery oblivion of a Big Crunch appears to offer less hope than a long, lingering journey into a freezing eternity of nothingness. At least such an eternity would offer the opportunity for something else to happen en route.

Whilst Vilenkin was working out the details of his 'universe from absolute nothing', in 1982, Stephen Hawking was working on his own ambitious theory, also based on a closed, collapsing universe from nothing. He called it a 'no boundary' model because it proposed that the universe was finite, with limits but no beginning – or end. *It was Hawking's model that provided the first inkling of the circularity of time.*

He based his model on quantum theory. It's all about scale. Where relativity describes the world at its largest, universe level, quantum theory describes the physics of the world at a subatomic level. What quantum theory has in common with relativity theory is its incomprehensibility. One of the greatest quantum physicists, Richard Feynman, was quoted as saying, 'If you think you understand quantum mechanics you don't understand quantum mechanics'. However, it is not necessary to fully understand quantum theory to understand the basic counterintuitive premise that at the subatomic level anything that can happen *does* happen and that a Big Bang, however it happens, must therefore always represent a multitude of possibilities. Hawking recognized that the Big Bang was the ultimate 'quantum state', and that meant lots of universes. His model of the universe is explained in the next chapter.

(Note: An examination of some of the more interesting and accessible aspects of quantum theory and its implications is included in Appendix B.)

CHAPTER THREE

Hawking, his universe and the first hint of circularity in time

'Oh there's so much I don't know about astrophysics. I wish I'd read that book by the wheelchair guy.' ***Homer Simpson***

When Stephen Hawking, Lucasian Professor of Mathematics at Cambridge University, discoverer of an 'impossible' property of cosmic black holes, and stubborn survivor of a form of motor neurone disease, wrote *A Brief History of Time*, neither he nor his publisher had any idea how well the book would be received. Five and a half million people bought the book in the first four years, placing it firmly on the best-seller list, a phenomenal record for a book of this kind, a book which, whilst intended by Hawking to be accessible to everyone, is by no means easy bedtime reading.

It is likely that the reason lies in the fact that in looking at the concept of time Hawking was attempting to offer a new insight into ourselves and our existence, as well as his chosen field. Hawking said that the aim of his book was to explain where the universe came from, how and why it all began, if it will come to an end and if so, how?

A Brief History of Time has undoubtedly been seen, therefore, as offering insight, not only into how the universe works, but into what it means for *us*.

Hawking's 'no boundary' model of the universe

A peculiar and unfortunate feature of general theory of relativity is that it actually breaks down at singularities, points of infinite density and zero size – it is not able to deal with the infinities it predicts. Unfortunately, not only relativity, but all the laws of physics break down at this point, making it not a little tricky to investigate. Stephen Hawking, now immersed in cosmology, was not at all happy with this, even though he was instrumental in discovering the singularity. He therefore turned to an approach more suited to the small-scale state of the universe at the beginning – quantum theory,

In *A Brief History of Time*, Stephen Hawking describes an elegant and highly original model of the history and future of a universe with no beginning and no end but which was not infinite. In so doing he became one of the first cosmologists to introduce the concept of circularity to the dimension of time. He showed that it is possible to have a 'finite universe', one with limits, yet no beginning or end. He called it a 'no-boundary' model. He assumed a universe with a Big Bang and a Big Crunch, but proposed that these momentous events were not the beginning or end of the universe. He derived his model from quantum theory and introduced a concept he called 'imaginary time'.

Hawking first presented his model in 1981 at a conference on cosmology organized by Jesuit priests at the

Vatican. They wanted to know how comfortably Christian orthodoxy sat with the latest developments in cosmology. Hawking's views on religion are such that he was bound to find the setting intriguing. In *A Brief History of Time* he recalls that his audience did not grasp the significance of his model, for a universe with no beginning presents problems for the Christian view of Creation.

Hawking's model has three cornerstones. The first and most important is the proposal that the universe could have begun from nothing at all, with all the positive matter–energy in the universe created from absolutely nothing – no matter–energy and no space – by creating at the same time an equal amount of negative gravitational energy. The condition that a universe could emerge from nothing is that it is closed. That is, it will eventually collapse into a Big Crunch. The energy conservation law dictates that no more matter can be created than can be 'reclaimed' from the gravitational energy created at the same time. If a finite amount of positive matter–energy is created then there will be a finite amount of negative gravitational energy and therefore a finite amount of expansion. For the universe to appear from nothing means that it must collapse into a Big Crunch at which not only matter–energy but also space itself, disappear.

Hawking's second cornerstone is that under the special conditions prevailing at its earliest stages, the universe must be considered a quantum system that is subject to the laws of quantum theory. In the simplest terms, this means that at its initial quantum state at the beginning of the Big Bang, not one, but a whole range of possible universe 'histories' lay open to it, including, of course, the one we are living in at the moment.

The third cornerstone, perhaps most fundamental of all, is the assumption that although relativity theory, the theory of gravity, appears to break down at the absolute beginning of time, t=0, it is the scale of time we use that breaks down, not the theory of relativity. Traditionally, indeed universally, time had been treated as a linear dimension with a possible beginning and an indeterminate end. Whilst there is debate as to whether it has a beginning at the Big Bang and even more debate as to whether it ends in a Big Crunch, there has been no question that it is a linear dimension. Hawking challenged that assumption. His solution to the breakdown of the relativity problem was to propose an alternative view of time. In this view, as soon as matter–energy appears in such prodigious quantities at the Big Bang in the unique conditions of the initial universe of minimal size and maximum gravity, relativity theory is not only relevant but is as important as quantum theory in determining future events. The two fundamental theories of science sit side-by-side holding dominion over the future of the universe.

He decided that the assumptions on which his singularity work with Penrose were based must have been wrong. Everything pointed in one direction. The singularities occurred at t=0, or 'time zero'. That is, the start of time itself. But what if it wasn't t=0 at the Big Bang? Why should the Big Bang be an absolute t=0? Did making t=0 effectively assume a singularity? Had he effectively only proved what he had assumed? There was only one conclusion: the conventional linear concept of time with a beginning, t=0, must be wrong. But, if so, what is the true dimension of time with no t=0? Relativity offered an answer.

Relativity theory shows that gravity is not an ordinary force; it acts by distorting space–time. The concept of space curved by gravity is now widely accepted. The notion of setting off in an absolute straight line into space and ending up where you started is not a figment of science fiction writers' imagination; it is a consequence of space–time curved round on itself by gravity in a closed universe. Less well known are the solutions of relativity theory first worked out by German mathematician Kurt Goedel, which showed the bending not only of space but also of the dimension of time itself.

Working in the early 1980s with James Hartle, a mathematician from the University of California, Hawking adopted this relativistic concept of a fundamentally curved dimension of space–time to which a Euclidean (flat) scale of measurement could be applied that would not disappear at the Big Bang; there would be no absolute beginning to time, t=0. This necessitated the use of imaginary numbers in the mathematics; that is, numbers based on the square root of minus 1. With this scale, time had no beginning, it had no end and it had no fundamental direction. This might have been expected. Relativity had shown space and time to be equivalent. Since space can have no beginning, end or direction, why not time?

Reflecting this creative use of imaginary numbers, Hawking called the timescale he devised 'imaginary time', to distinguish it from the conventional linear concept, which he called 'real time'. Recognizing that the idea of 'real' and 'imaginary' times was confusing, he said:

> 'This might suggest that the so-called imaginary time is really the real time and that what we call real time is just a figment of our imagination ... maybe what we call imaginary time is really more basic and what we call real time is just an idea that we invent to help us describe what we think the universe is really like.' (1)

The word 'imaginary' was unfortunate because it gave the impression that it was a mere figment of his imagination, a trick. As a result, his work was condemned by some scientists. Hawking recognizes the problem in *Black Holes and Baby Universes*, in which he commented:

> 'With hindsight I now feel that I should have put more effort into explaining ... particularly imaginary time, which seems to be the thing in the book with which most people have the most trouble'. (2)

In fact, the use of imaginary numbers has a well-established precedent. They are used in alternating current theory. The circularity at the heart of alternating current theory (the circular motion of the generator that produces alternating current) creates mathematical difficulties that are considerably simplified by using imaginary numbers. No one has suggested that this is in any way unscientific. Yet this criticism was levelled at Hawking. Perhaps it was because it was not sufficiently clear that he was assuming circularity in the timescale he was dealing with.

Using imaginary time in the mathematical approach they had developed, Hawking and Hartle set about analysing the whole universe as a quantum system of matter–energy created from nothing, dominated by gravity. The simplicity and elegance of the result was reassuring. The quantum possibilities arising from that initial state – the allowed histories of the universe – could be represented by a simple space–time model. In many ways it was similar to the conventional model, expanding from the Big Bang and contracting to the Big Crunch. The difference was that, using a different timescale, these two events are not a beginning or end.

Hawking's model is difficult to get to grips with, partly because imaginary time is not fully explained and partly because we are resistant to readjusting our natural way of looking at time. It helps to start by reconsidering the conventional Big Bang theory.

The conventional model of a closed universe

The important thing to remember about the conventional view of the evolution of a closed (expanding then collapsing) universe is that the dimension of time used is linear, with an absolute start, time $t=0$, at the Big Bang, an absolute end at the Big Crunch, time $t=E$, and an absolute direction from $t=0$ to $t=E$. We really have no idea what E is, except that it is in the order of hundreds of billions of years. The conventional closed-universe model (Fig 3.1) is essentially Friedmann's first model. The graph shows the separation of galaxies plotted against the linear dimension of time. The separation increases from zero at the Big Bang up to a maximum and then reduces again to zero at the Big Crunch. If, instead of galaxy separation, we use the spatial size of the expanding universe and extend this below the time axis, we can represent the spatial size as circles (Fig 3.2).

In this model, the spatial size 's' of the universe is represented two-dimensionally by a circle, starting from nothing 's=0' at the Big Bang 'B' where t=0 (time equals zero) and collapses to nothing again 's=0' at the Big Crunch 'C' where 't=E' (End of time). It reaches its maximum size at t=E/2 – that is, half way through. The feature of this model is that it has distinct points at t=0 and t=E, marking them out clearly as fundamentally special points, the 'singularities' mentioned earlier at which the laws of physics break down. With this model, our natural inclination is to see time as flowing between these limits, beginning and end, with the life of the universe rather like the life of a meteor that appears at the Big Bang, displays its bright trail of existence and then disappears forever at the Big Crunch. In the conventional closed-universe model, the Big Bang and Big Crunch are unbreachable boundaries, the beginning and end of the universe.

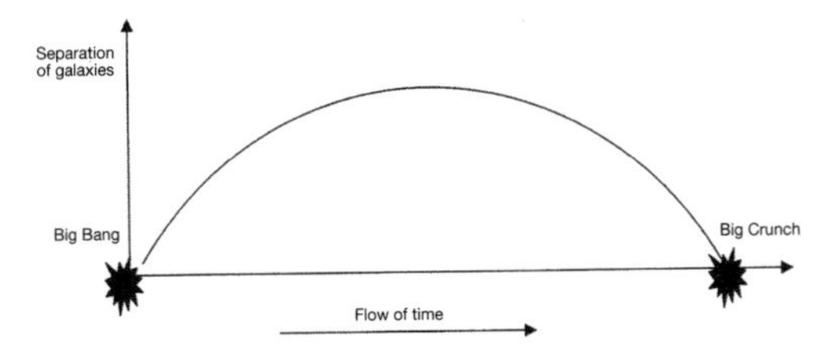

Fig 3.1 Friedmann's graph showing expansion and collapse of a closed universe

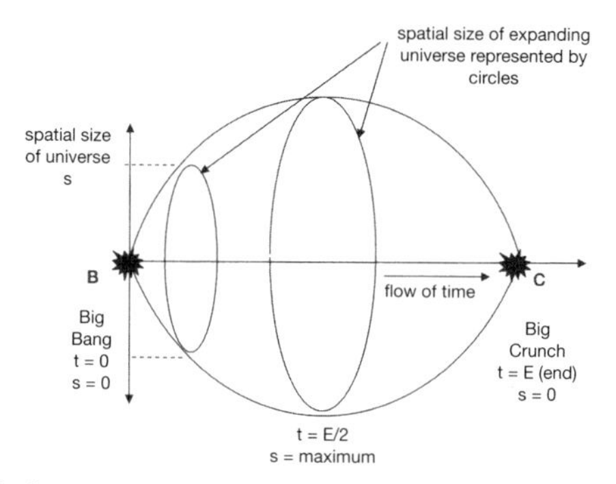

Fig 3.2 Friedmann's graph converted to a space–time model of a closed universe

Both the origin and fate of time in this universe are unclear. It has been accepted that, in some way, time as well as space begins and ends at the Big Bang and Big Crunch, though there has remained the obvious and tempting speculation that there is somehow a more fundamental dimension of time along which the Big Bang and Big Crunch are but two events.

Hawking's model of curved space–time

Recall that, using a quantum theory perspective, anything that can happen will happen. Hawking proposed that at the Big Bang there must have been not one but a set of quantum possibilities for the future of the universe and that it is impossible to predict which of these possibilities would emerge. In the strange world of quantum theory, the only way to understand these possibilities is either to consider them all hovering in some kind of 'probability limbo' until one – ours, with us in it – crystallizes out into reality according to its probability, or to imagine them all appearing together as 'parallel universes'. Hawking favoured the latter, though as he remarked:

> 'Exactly what meaning can be attached to the other histories in which we do not exist, is not clear.' (3)

Using their own scale of imaginary time, Hawking and Hartle created models of the quantum possibilities that could arise from the initial Big Bang quantum state, incorporating into each possibility all the essential characteristics of the conventional model – a universe with zero size at the Big Bang, an expansion to maximum size and collapse back to zero size at the Big Crunch.

It might have been expected that each quantum possibility would require its own model. However, Hawking and Hartle showed that all the quantum possibilities had the same characteristics and could therefore all be represented using a single visual model. Even better, it could be represented by a subtle reworking of the conventional model. The resulting space–time model is shown in Fig 3.3. The model retains the principle of representing the size of the universe by circles, and therefore looks at first sight much the same as the conventional model. The universe expands from nothing, creating its matter–energy out of gravitational energy, reaches a maximum size and collapses back into nothing once more at the Big Crunch. The difference, however, is that where the conventional model uses the straight line between the Big Bang 'B' and the Big Crunch 'C' as the dimension of linear time, with a fundamental direction from 'B' to 'C', Hawking represents time – 'imaginary' time – along the *surface* of the model and it has, like the dimensions of space, no fundamental direction. Thus, where the conventional model shows distinctive points – discontinuities – at the Big Bang and Big Crunch, Hawking's model, being perfectly spherical, has no discontinuities at 'B' and 'C'. This means that since time is represented by this surface, there is nothing special about time at 'B' and 'C'. The Big Bang and Big Crunch are no longer special points on this curved surface of imaginary time as they were on a linear dimension of time.

To help us get our heads around this strange proposal, Hawking likened his model to the earth. 'B' and 'C' are like the North and South Poles on the earth (Fig 3.4). These poles appear to be special points on the earth's surface but in fact they are simply the ends of a scale of measurement used for navigation, based on the earth's axis and its magnetic field. The earth's surface itself isn't different there and it doesn't end there. Though the latitude scale imparts a sense of direction to the surface of the earth, with the poles as beginning and end, it is the latitude scale that has the direction, beginning and end, not the surface of the earth it measures. There is no beginning and end to the surface of the earth. And whilst this surface provides a dimension across which movement can occur, it doesn't move itself. We don't talk about space 'flowing', 'passing' or 'marching on' as we do about time.

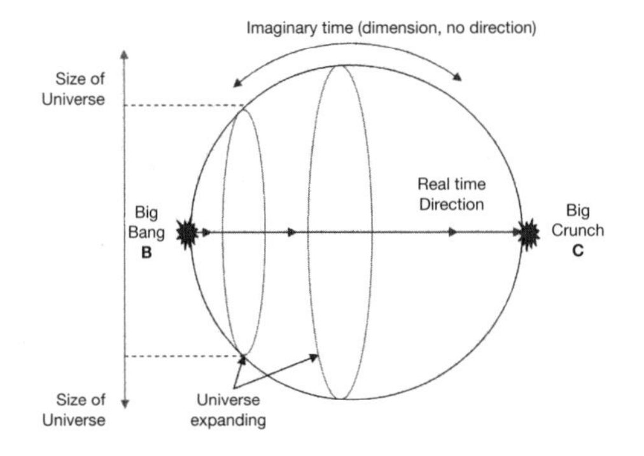

Fig 3.3 Hawking's model of the expanding universe showing real time as the conventional time axis with a direction starting at the Big Bang B and ending at the Big Crunch C. Imaginary time is the dimension over the surface with no direction and no beginnning or end at the Big Bang, Big Crunch or anywhere else.

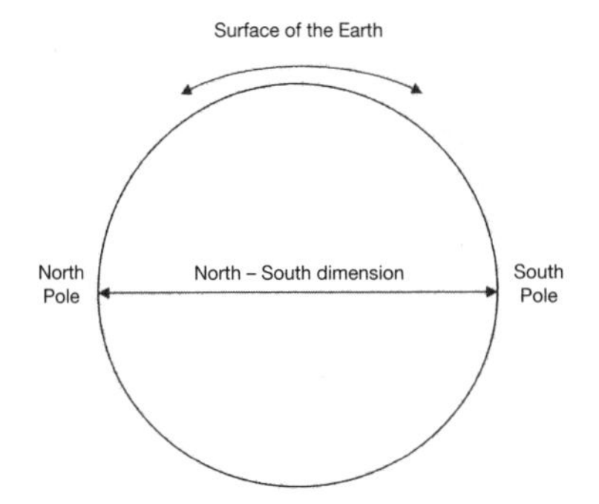

Fig 3.4 The analogy of the earth and the North-South dimension to illustrate imaginary time. Just as the earth's surface does not start and end at the North and South Poles, so imaginary time does not start and end at the Big Bang and Big Crunch.

Hawking did not argue that imaginary time was a more accurate description of time than the conventional view of time. He presented it merely as an alternative. However, if imaginary time exists, it means that time is not what conventionally we understand it to be; we have to accept a sense in which time is curved like the dimension of the surface of the earth so that it has no beginning or end. And being only a dimension it has no fundamental direction; it doesn't 'flow' or 'pass' or 'march on'. Time cannot 'run backwards' for the simple reason that it cannot run anywhere. Just as the earth's surface sits there with places scattered over it, so the dimension of time just sits there with events scattered over it. Most importantly, the Big Bang and Big Crunch are merely two events on the dimension of time like the poles of the earth. They are not the beginning and end of time, any more than the poles are the beginning and end of the earth, merely the beginning and end of a convenient scale we use to measure time – the conventional, linear time-scale that Hawking called 'real' time. However, Hawking is suggesting that it is still merely a scale, not the dimension of time itself.

Nevertheless, just as the poles represent the extent of the earth – what is between them is all there is – so the Big Bang and Big Crunch represent the extent of the universe in space and time. The space–time between them is all there is. And like the surface of the earth, the whole of space–time in a sense already exists; the future of the universe, along with its past, is already mapped out. We, along with all the other events, simply move through it like explorers walking through uncharted jungle.

Where does our sense of time flow come from?

Why does the idea of time starting at the Big Bang and flowing towards the Big Crunch with all the events of history, including ourselves, in-between feel so natural to us, and why does Hawking's idea of it 'all being there already' feel so unnatural? The answer is that there is a connection between all these events along the dimension of time that gives us a powerful sense of direction, beginning, end and flow. And this happens in exactly the same way that we get a sense of direction, beginning, end and flow from a row of milestones.

To a child, milestones might appear to be simply lumps of stone scattered along a road. As soon as their significance – the connection between them – is understood, however, they suddenly represent direction and movement. They show how close we are getting to a particular destination. To us, events are like milestones, in terms of time direction and progression, because we are 'programmed' to see the connection between them. This time direction is, of course, the direction in which we get older, in which we remember things in the past but not in the future. This is not simply psychological, however. It is more fundamental than that. Our bodies – and minds – work thermodynamically. Indeed, everything that happens does so by thermodynamic processes. That is to say, processes initiated by the movement of heat and the associated energy changes. The history of the earth is the history of thermodynamic processes – the processes set in train by the sun's heat as its nuclear furnace burns away. These processes always occur in one time 'direction' – the direction in which heat travels naturally from hot to cold. All events in the universe occur in this way. Since we are part of this thermodynamic process – all human body processes, including our memory processes, are thermodynamic processes – our psychological direction of time is in the same direction. It is fundamental because it is the direction in which the universe as a whole is cooling down, and it is this cooling that gives rise to all the events in the universe, including our lives and the way we think. If the temperature across the whole of the universe equalized as a result of this cooling, there would be no more *events* and no more *us* to observe them. This is the 'heat death' of the universe. The time direction in which the universe is cooling down is the direction in which it is expanding from the Big Bang. Since both we, and the events we observe, are driven through time by thermodynamics, we could be forgiven for conceiving of the dimension of time itself as flowing, with a real direction.

We could be forgiven, says Hawking, but we would be wrong. All this flow, direction, beginning and end of time are subjective. We see things in this way because we are part of the thermodynamic process. If we weren't, things would seem very different. Once again, we meet an important principle – that we are part of the natural processes we are investigating. We would love to assume that we are in some way above it all and can therefore look down, rather like God, quite independent of what we are observing. In fact, only if we accept that we cannot examine everything as independent observers can we truly understand the world and our place in it.

We can get a nice idea of our inescapable thermodynamic perspective by thinking of the earth again with its magnetic field. Magnets as we know always have two poles. If we could make such a thing as a single magnetic South Pole it would tend to travel from North Pole to South Pole and stop there for good. If this magnetic pole could think, it would assume that the earth's surface has a fundamental direction, with an absolute beginning at the North Pole and an absolute end at the South Pole. But this would simply be its 'subjective view' of the earth, the result of its magnetic nature. In Hawking's view we are like this magnetic pole, but we are thermodynamic not magnetic. Our thermodynamic perspective gives the illusion of time flow and an absolute direction to time in the same way that the magnetic pole would have the illusion of flow and an absolute direction to the earth's surface. Time is simply the dimension through which, thermodynamically, we 'move' and all other events occur. Time is the road, events are the milestones and we, along with other thermodynamic reactions, are the travellers.

Another way of looking at it is to think of one of those advertising hoardings with a row of bulbs that flash in sequence, making the light appear to travel along a board. The row of bulbs doesn't move; nothing moves. There is simply a set of connected events – the flashes – occurring on the board that, because of the sequence in which they light up, give the appearance of movement. The message we get from Hawking's model is that we must consider that time, the fundamental dimension, is like the stationary board across which events in the history of the universe occur like the flashing lights. Our conventional flowing, directional concept of time – Hawking's real time – is then the apparent flow of these events across this board of time. Like the movement of the light pattern, the flow of time is illusory, the result of our interpretation of the connection between the events that occur.

This doesn't mean that our conventional view of time is wrong. It means that it is an analogue of time, not time itself. It follows time accurately and so we can use it to measure time and to make calculations. It must be put in the same category as a clock, which is a perfectly accurate analogue of time, and therefore a very useful one, but it is only an analogue. It is not the dimension of time itself.

So we are left with Hawking's view of time as a finite, self-contained dimension with no beginning or end, like the surface of the earth. It is not a flow of anything nor has it an intrinsic direction. Recall the tale of the lost motorist who asks an old farmer, 'Does this road go to London?' 'The road don't be goin' nowhere,' comes the reply. 'If you want to get to London you'll have to use that there car of yours.' Like the road, time just sits there with events like milestones scattered along it. The whole dimension of time 'exists already', with the events of the future waiting for us to arrive at them. Along this dimension there is a particular series of events that have a special meaning to us because they are ordered in exactly the same time direction as we are – thermodynamically. We are programmed to detect time only by the events that occur along it, like the ticking of a clock, the movement of the earth around the sun, or the appearance of wrinkles and grey hair when we look at ourselves in the mirror. As a result, the 'path' traced out by these special events has become our 'real' time because, like the path of the magnetic pole from north to south, it is the only path through time that has any real meaning to us.

Since the universe appears at the Big Bang and disappears at the Big Crunch, these two points in the thermodynamic process lend themselves perfectly to being the beginning and end of our chosen scale of measurement of time. Our sense of time flow is simply our natural perception of this course of thermodynamically connected events. This thermodynamic scale of time – 'real' time – serves us perfectly well in most situations, just as Newtonian mechanics does. But like Newtonian mechanics, it fails us when we are working on a cosmological scale, when we want to know what time *is*, rather than how events occur along its dimension.

But what *is* imaginary time?

Hawking's 'real' time is so-called because it has a very real physical significance. The problem is that Hawking's model was based on his 'imaginary time'. We might reasonably ask, what is its status as the basis of a model of the universe? Does it have any physical significance, like 'real' time? What if imaginary time *is* no more than a trick? Does everything else go out of the window with it? In fact, Hawking acknowledged that imaginary time offers no insight into our existence because we live in real time. If we could continue to the Big Crunch we would still be crunched into oblivion. Hawking admits that even in his universe:

> '...in a sense we are still all doomed.' (4)

This model seemed to hold out such high hopes for a much deeper understanding of the universe, of time and of our existence. But Hawking's own conclusion suggested this hope had been dashed. However, a moment's thought tells us that there is a ray of hope, for it cries out for the answer to a very simple question:

What is the difference between the two space–time points, the Big Bang and the Big Crunch? In Chapter 4 we shall look at the answer to this question and its important implications.

CHAPTER FOUR

From imaginary time to circular time – the Principle of Identity

'I got back to my house tonight and found that everything had been taken, and replaced with exact copies.' ***Steven Wright***

The significance of 'nothing'

As in the conventional expansion–contraction model, Hawking identified the Big Bang and Big Crunch as different events in space–time, even though he accepted that the universe starts out from nothing and ends up as nothing. Like the North and South Poles on the earth, in his model the Big Bang and Big Crunch are very different points at opposite ends of the dimension of time. He did get as far as considering the possibility that the contraction phase of the universe was a mirror image of the expansion, and even the possibility that the thermodynamic direction of time reversed during the contraction phase; that is, real time would go in reverse, so that broken cups would reassemble as in a video played backwards. But he quickly dismissed the notion. Yet, he did not consider one possibility – that the Big Bang 'nothing' could be the same space–time point as the Big Crunch 'nothing'.

By definition there is nothing by which to discern any difference between two 'nothings'. On what basis can we assume they are different points? There can be no more appropriate case for the application of the Principle of Identity, which is central to this book. This principle is derived from Leibniz's *Identity of Indiscernibles,* which says that *no two distinct things exactly resemble one another.* The corollary is the Principle of Identity: *when two things, events or series of events are absolutely indistinguishable, they must be regarded as one and the same.*

This principle, though intuitive, is so important that it's worth saying more about it. If we went for a walk and came to a place that was indistinguishable from a place we had passed earlier, we would assume without a second thought that we had walked in a circle. We would not consider for a moment that we had come to a different, though identical, place. We are not surprised, when Portuguese explorer Magellan's expedition ended up where it started in the 1500s, that the crew assumed their voyage was circular, rather than that they had arrived at an identical place in a different location. We saw earlier that Einstein based his theory of general relativity on the realization that it would be impossible for anyone to distinguish between the effect of gravity on earth and the effect of being accelerated at 'g' (10 m/33 ft per second per second) in gravity-free outer space, or between free-falling on earth and floating in outer space. He used this to show that gravity is not an ordinary force but the result of space being bent by matter. These are all examples of the Principle of Identity at work.

In fact, the principle is much more fundamental than we realize. It is applied by everyone virtually every second of every waking moment, with its use passing unnoticed. It is the basis of recognition. When we come home from work and greet our loved ones, we assume they are the same loved ones that we left when we went to work. We wouldn't consider the idea that they could somehow have been replaced by identical copies.

The question is, then, in what way is the 'nothing' from which the Big Bang emerges different from the 'nothing' into which the universe collapses in the Big Crunch? We know, of course, that the *departure* from the Big Bang is different from the *arrival* at the Big Crunch, but that in itself doesn't make them different points. The arrival at a train station is quite different from the departure but we don't say that there must therefore be two stations.

Consider the meaning of 'nothing'. Not the nothing of empty space, which is the vacuum left when all the matter–energy has been removed, but *absolute* nothing, no matter or space. There is no conceivable way in which one such

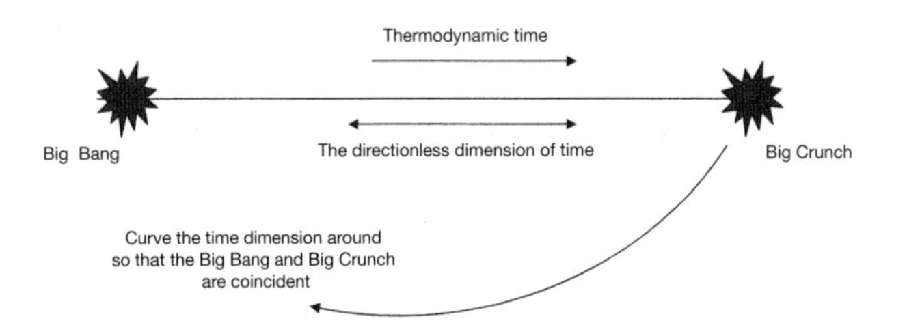

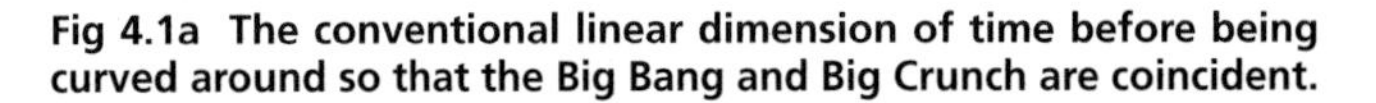
Fig 4.1a The conventional linear dimension of time before being curved around so that the Big Bang and Big Crunch are coincident.

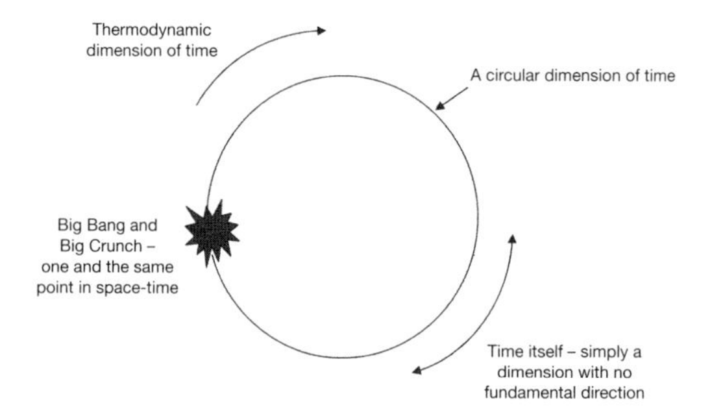

Fig 4.1b The circular dimension of time with the Big Bang and Big Crunch as one and the same event in space–time. Like space, time has no fundamental direction, but the evolutionary events of the universe, of which we are one, follow the thermodynamic directions from Big Bang to Big Crunch.

'nothing' can be different from another. By definition, there is absolutely nothing to *be* different. Even if, as Hawking did, we consider the 'nothing' to be a quantum state, the quantum possibilities at the Big Bang 'nothing' must be identical to those at the Big Crunch 'nothing'. Indeed, his model seemed to specifically remove any distinguishing features of the Big Bang or Big Crunch points. Whichever way we look at it, we see only one conclusion: it is impossible to distinguish between the states of the universe at the start of the Big Bang and the end of the Big Crunch.

According to the Principle of Identity, this means that the Big Bang and Big Crunch must be considered as one and the same point in space–time. When the universe arrives 'thermodynamically' at the Big Crunch it has actually arrived back where it started at the Big Bang, from which it must then begin all over again towards the Big Crunch, and so on, ad infinitum.

This leads to a profound conclusion: the dimension of time is not merely curved; it is completely circular. Fig 4.1a shows the conventional linear dimension of time with the Big Bang and Big Crunch at its beginning and end. If the Big Bang and Big Crunch are one and the same point, we must curve this linear dimension around to make them so. This gives us the circular dimension shown in Fig 4.1b.

No beginning, no end but no eternity?

The proposal of a universe from nothing removed the problem of the idea of the universe emerging from a singularity of infinite density and zero size. A completely circular dimension of time, with the Big Bang and Big Crunch coincident, takes us a step further.

It embraces Hawking's simplifying concept of a finite universe with no beginning or end, of a dimension of time

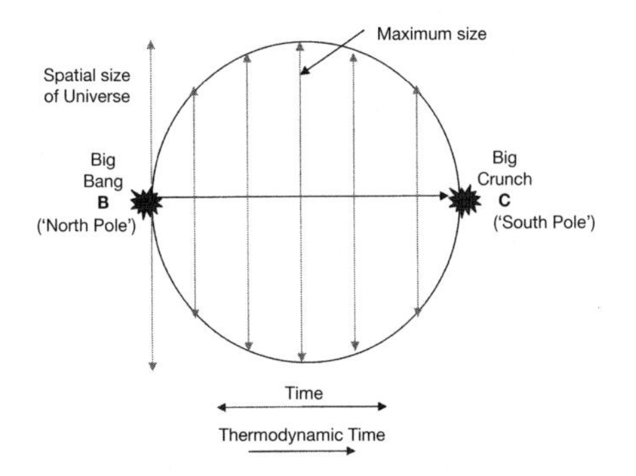

Fig 4.2 Hawking's model in 2D – from a sphere to a circle.

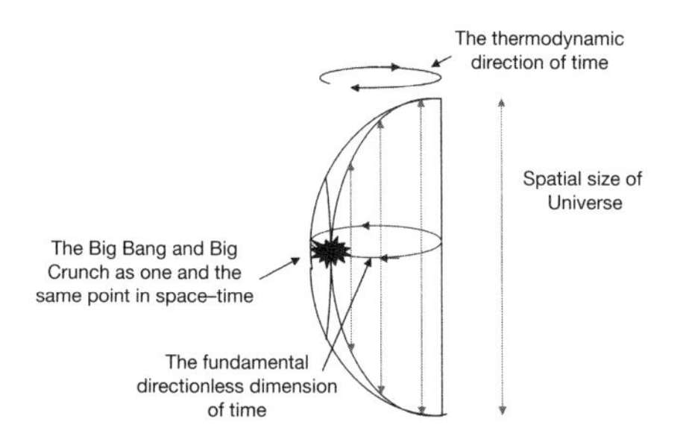

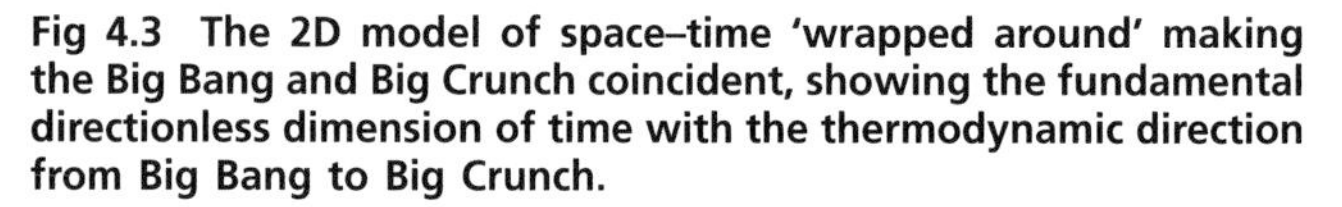
Fig 4.3 The 2D model of space–time 'wrapped around' making the Big Bang and Big Crunch coincident, showing the fundamental directionless dimension of time with the thermodynamic direction from Big Bang to Big Crunch.

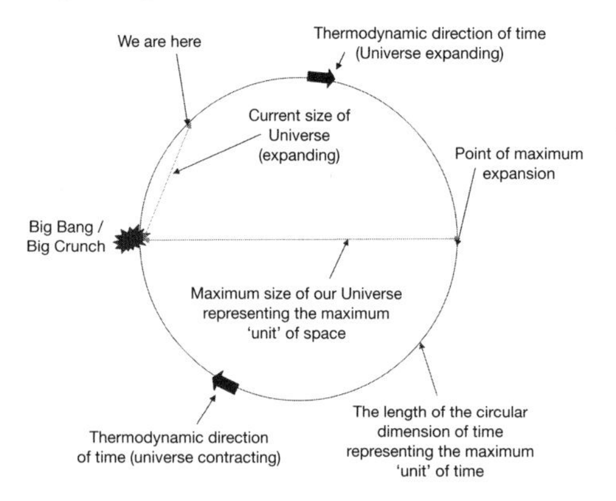

Fig 4.4 The simplest space–time model of the universe showing the circular time dimension and thermodynamic direction, together with the size of the universe.

that, like space, has no fundamental direction, but without the need to introduce an alternative, arguably obscure scale of time to explain what happens at the Big Bang or Big Crunch. Everything, the universe, space and time, is contained in an endless circle of time. So, if for the moment we dispense with Hawking's idea of using the outer edge of the model as the curved dimension of imaginary time, his North/South Pole model in Fig 4.2 changes to Fig 4.3.

We no longer have to use an edge or surface to show a circular dimension. The thermodynamic direction simply goes from B round to C and continues around forever.

The 'wrap-around' model in Fig 4.3 represents all the essential features of Hawking's spherical no-boundary model – no beginning or end, no fundamental direction, and containing all that exists in space and time. There is nothing we can detect 'outside' or beyond it. We cannot at the moment estimate the length of a single cycle from Big Bang to Big Crunch. Our universe appears to be so near the point of initial expansion that any estimate can be no more than speculation. If our universe is about 13.7 billion years old it would be reasonable to suggest a cycle time of at least 100 billion years.

The history of the universe is then the chain of thermodynamic events contained within a single cycle of expansion and contraction as endless but as finite as an elastic band. Fig 4.3 can be further simplified, offering the simplest possible space–time model – a circle of time whose diameter represents the maximum expansion, and therefore size, of the universe. (Fig 4.4). Our life histories are events along that circle of time, destined to recur in subsequent cycles. If the model in Fig 4.4 is correct, there are interesting implications for the ultimate 'Theory of Everything'. Whatever the theory turns out to be, however complex or diabolical its equations, a circular dimension of time demands that the theory must predict a universe expanding and contracting within a circular dimension of time. It is a framework that satisfies Hawking's own condition for the ultimate theory – that it will be 'understandable in broad principle by everyone, not just scientists' (1).

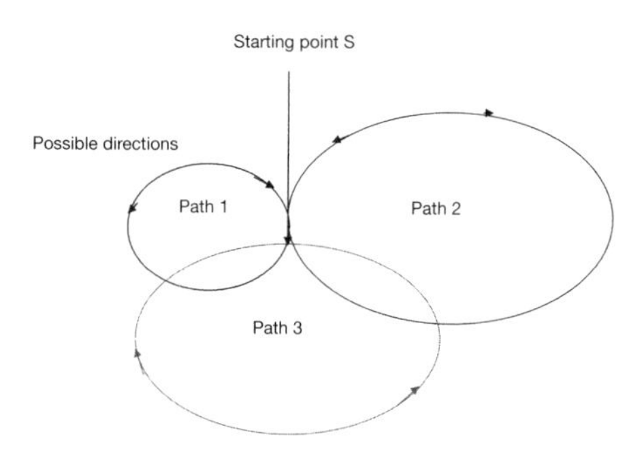

Fig 4.5a Alternative paths from a single starting point.

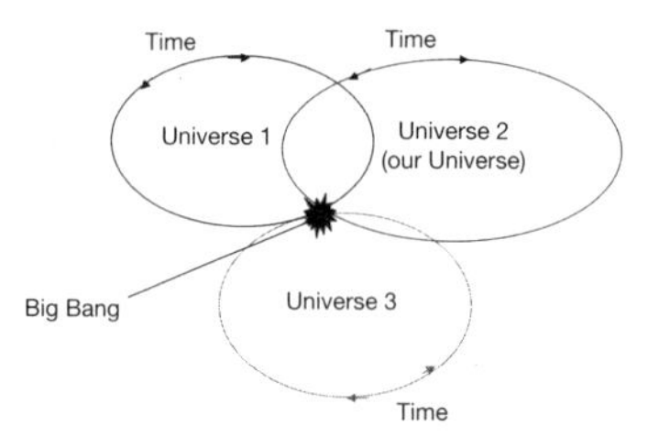

Fig 4.5b Three possible universal time cycles emerging from the Big Bang, each with a characteristic probability of existence.

Here we meet a slight complication, and it comes from quantum theory, from which, remember, Hawking derived his model. Quantum theory tells us that the Big Bang is a quantum state from which not one but a whole range of universes must emerge, so that every time we arrive at the Big Crunch we find ourselves back at the Big Bang with all those quantum possibilities available once more, including the one we are now living in. This sounds ominous. It looks as if the neat conclusion we arrived at is not going to be so neat. The good news is that for our purposes all but one of those different universes can be ignored since none but that one will be *our* universe, with *us* in it. Yet we can see some meaning in those others.

Hawking's way is to imagine all the possibilities occurring somehow together as 'parallel worlds' so that in a sense all the other universes somehow exist to emerge at each Big Bang as different and inaccessible dimensions. We can see this if we imagine the different ways of walking along different paths, from a single starting point and back again.

Fig 4.5a shows three such paths. Notice, incidentally, that in this path of analogy, though there is a direction to our travel, the paths themselves have no direction, neatly reminding us that while we travel through time in the thermodynamic direction, time itself, like the paths, has no fundamental direction. If we were to explore all these paths in order, 1 to 3, each time starting from S, then from our point of view as travellers, these paths appear one after the other. However, we know that all these paths exist already. So are these paths really one after the other or do they all exist together like 'parallel worlds'? The answer, of course, is both. This precisely mirrors time and the quantum possibilities at the Big Bang as shown in Fig 4.5b. From the Big Bang there emerges an unknown number of possible universes, one of which must be the one we are now living in. The others are the 'multiverses' much beloved by sci-fi adherents and now firmly established in cosmology.

A quantum pinball machine

The notion of the universe starting out, going through a history of events too complex to predict but then returning to its starting point, is reminiscent of a pinball machine. Each pinball game, like each cycle of the universe, is identical in starting and finishing positions but different in detail. There exists, for the ball at its starting point, a whole range of possible pathways – histories – each having a certain probability. It is not possible to predict which history will occur, but whichever does occur, the end point will always be the same. The ball will always return to its initial position. So, if we play the machine often enough, no matter how complex the path of the ball, no matter that this is a classic example of chaos theory in which the tiniest variation at any point in the ball's path will transform its subsequent behaviour, we will eventually see the ball taking an identical path to one it took earlier. Apart from the fact that the beginning and end of the pinball cycle are at the same point only in space, rather than space–time, this mirrors our model of the universe. We might try to predict the path – the 'history' – of a ball through its cycle by working out the equations governing the multiplicity of impacts and spin variations experienced by the ball in its path through the machine. But even if we could predict each individual path of the ball, our results would simply show that the ball ends up where it started. As we congratulated ourselves, the little boy on the pinball machine would tug at our sleeve and point out that he could have told us where the ball would end up without going to all that trouble. The point is that just as we don't need to understand the structure of the earth to predict its motion around the sun, we may not need to understand how the universe works at the quantum level to understand how it works at the cosmological level.

This may have a message for us in our search for the 'Theory of Everything'. If we do eventually discover those ultimate equations that constitute the 'Theory of Everything', then just as the ball in the pinball machine ends up where it started, we may simply discover that the universe starts out as nothing and ends as nothing. The equations that govern each and every field of physics might simply distil down to one equation that represents the separation of nothing into matter–energy, m, and gravitational energy, E:

$$\mathbf{mc^2 - E = 0}$$

Multiverses may lie in extra dimensions of time

One of the features of the theory of relativity was that it showed that time was not a separate dimension, independent of space. Space and time are inexorably linked in the dimension of space–time. Since space can be curved by gravity it should not seem surprising if time were curved, too. However, the existence of the Big Bang quantum histories offers a further equivalence between space and time. The conventional view of space–time is of three dimensions of space (height, breadth and depth) and one dimension of time (linear). But these four dimensions do not accommodate the quantum history possibilities. The other quantum histories, if they are considered at all, are generally seen as somehow floating around in some hyperspace totally disconnected from our universe. Yet they are connected – at the Big Bang. We might ask, therefore, if time, like space, could have more than one dimension in order to accommodate these quantum possibilities with the Big Bang as an event common to all of them. We know that these possible histories must exist in a different dimension because they are inaccessible. Mathematically, these inaccessible 'parallel universes' would not be parallel at all but would exist in another time dimension at *right angles* to the first, like the space dimensions. This theory is discussed in Appendix A. Here we are interested in the implication *for us,* of a circular dimension of our universe in what Stephen Hawking called 'real time' but what we understand simply as 'time'.

CHAPTER
FIVE

From the circle of time to the circle of existence

'Millions long for immortality who do not know what to do with themselves on a rainy Sunday afternoon.' ***Susan Ertz***

The concept of a circular dimension of time along which the history of the universe unfolds like a circular bus route (Fig 5.1) always returning to the same place in space–time, is remarkable enough. But it implies something that is of greater importance – a theory of our existence, one that tells us the answer to the question that started this book – what exactly happens to us when we die?

The implication is unavoidable:

If the history of the universe is repeated precisely in every detail from every Big Bang/Crunch point then our lives, hitched as they are to the universe, must also be repeated in the same precise detail. When we die we must be reborn in the 'next' universe exactly as we were in this life.

We have arrived at this conclusion simply by applying the Principle of Identity to the two states of absolute nothingness at the Big Bang and Big Crunch that are the limits of a closed universe. It could hardly be simpler or more persuasive. Quantum theory complicates the picture a little, but affirms the underlying picture. In a closed (expanding then collapsing), universe, quantum theory proposes that a multitude of possible universes must emerge from the quantum state of the Big Bang, best imagined emerging all at once as multiverses. One of these must be a repeat of our current universe, repeated in every detail, not simply a universe 'a bit like this one'. This is ensured by a circular dimension of time. And since our existence is part of this universe, there must also be a repeat of our existence. And it must be a repeat in precise detail because, as our space–time model makes clear, it is actually the *same* universe. We have simply returned to the start of 'our universe'.

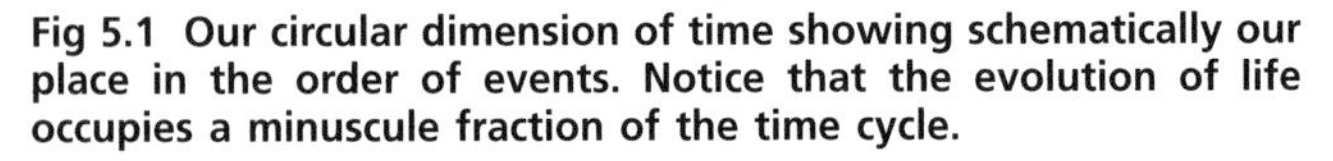

Fig 5.1 Our circular dimension of time showing schematically our place in the order of events. Notice that the evolution of life occupies a minuscule fraction of the time cycle.

All we need to know is that in a closed universe our lifetimes will be part of an endless but finite circle of time in which, when we die, we have the prospect of being born and living our lives all over again, precisely as before, in a never-ending circle of time. It is crucial to keep in mind that when I refer to 'the 'next' cycle, it is *this* cycle revisited. That's why our lives recur precisely. There are no 'future' cycles, any more than there are 'future' circumferences of the earth for us to travel around. Nevertheless, the circularity of time means that for us the offer of life after death is no longer exclusive to religion; it is now on offer from science, albeit a 'mere' repeat of our current life.

Patience is a virtue

Remarkable though this is, it leaves us with a problem. The cosmology of a closed universe dictates that the time from our death to our rebirth in the 'next' cycle will be immense, virtually the length of the whole cycle from Big Bang to Big Crunch. We really don't know how long that is. If our universe has been around for nearly 14 billion years since the Big Bang then it is reasonable to suppose that the whole cycle is of the order of hundreds of billion years. This would appear to be the time we would have to wait after our death until we reappear in the 'next' cycle. This unimaginably immense passage of time could seem as good as never. It would certainly present a problem for us if we had to experience it.

But we don't.

Anaesthesia not only stops pain, it stops time – and so does death, but more so

We can only experience time when we are conscious. Time can only exist for us if we are aware of events occurring in time. And we can only do this if we are conscious.

You can get an idea of this if you think of the passage

of time when you are asleep. If you sleep really soundly you are not really aware of the hours you are asleep, even though you are actually still conscious to a degree, enough to roll over many times and perhaps to dream. A better example of complete unconsciousness is seen in anaesthesia. If you have ever been given a deep general anaesthetic in hospital you will know that it seems to you that at the very instant you become unconscious you are awake again. You have no perception of time passing whilst unconscious for the simple reason that consciousness is essential to experience the passage of time. It is therefore common to wonder if the operation has actually been done.

But what about the ultimate in unconsciousness – death? Even under an anaesthetic or even in a coma, key bodily functions are still working, so it is conceivable that your brain could somehow sense the passage of time, even if you are not aware of it. But in death absolutely nothing is working; nothing to register time even by the activity of the cells in your body. From the moment you die, those zillions of years later when you re-appear in the 'next' universe, will effectively not exist. As far as you are aware, the instant you die you are reborn as yourself with precisely the same future ahead of you as you had in this life.

We therefore have a thoroughly rational, scientific possibility – indeed inevitability – of life after death. We won't know anything about it in this 'next' life because at our re-birth we shall have the same memories as we had when we were born in this life – absolutely none. All we can say is that if we were watched by a cosmological fly-on-the-wall, it would see our life unfolding in the 'next' universe in precisely the way it unfolded in this life.

A niggling problem may remain. In this repeat life scenario, will the reborn 'you' in the 'next' life actually be you and more importantly, *feel* like you? The answer was given above. Because the 'next' universe is this universe repeated, the 'you' in it must *be* you and therefore feel like you, living your identical life.

Perhaps before we continue we should just consider, briefly, the possibility that some people may think all this gives scientific legitimacy to the notion of reincarnation.

Reincarnation and past life regression

A repeated life does sound rather like reincarnation. Reincarnation is, however, a quite different phenomenon. It involves returning as someone, or something else, at different points in space–time. A circular dimension of time means simply returning as ourselves when our universe comes round again; a future universe in circular time is still *this* universe. The circularity means that we shall have returned to exactly the same period in space–time, just as when explorer Ferdinand Magellan's crew arrived home they were returning to exactly the same point in space. A 'past' existence in circular time is not really in the past; it is just as much in the future as in the past, so we can have no memory of it. Memories of 'previous existences' as *someone else* – 'past life regression' – are similar. They cannot be memories of a previous universe. Scientific consensus is that if they are memories at all, they are memories from *this* universe or, more likely, a combination of such memories with a healthy imagination.

CHAPTER SIX

Storm clouds brewing for the closed universe and the Big Crunch

'Only two things are infinite – the universe and human stupidity – and I'm not sure about the former.' ***Albert Einstein***

Everything that has been said up to now is based on a closed, collapsing universe, a model stemming more from theoretical physics rather than from empirical research. That was to change at the turn of the century.

From the very beginning, gravity has been acting to slow down the expansion of the universe, trying to pull it back together again. If it were to succeed, as in the first of Friedmann's possibilities that we have just been looking at, then not only would the universe be 'closed', ultimately collapsing into a 'Big Crunch', but it would also determine the actual shape of the universe – space itself – until that happened. Space would resemble the earth in that if you set off in a straight line you would end up where you started. This would also happen to a beam of light which, even though it travelled in a straight line across the universe, would end up where it started because of the shape of space. And if we could look far enough into space we would see the back of our head (we wouldn't, of course, because we would be long dead before the light from the back of our head reached us!).

We have seen that in this closed universe scenario not only would space be circular, albeit with such a huge radius that it would seem to be absolutely flat, but so would time, with the profound implications we have seen for the universe and for us.

Until recently, it was believed that if there is any such slowing down in the expansion of the universe it is so slight as to be undetectable. The universe would not collapse for a very long time. Recall the theoretical possibilities that came from Einstein's relativity theory. Two came from Friedmann – the closed, collapsing universe we have just been looking at and an open, ever-expanding, infinite universe. The third was a variation of the second in that the expansion of the universe would slow down, always on the brink of collapse but never actually collapsing into a Big Crunch. That would mean that it was flat, as opposed to the circularity of a closed universe. A beam of light would continue to travel in a straight line forever.

The universe has been expanding for so long – at least 13.7 billion years – that its shape is inevitably going to appear to be flat, just as the earth looks flat to someone on its surface. Until recently, the information we have had was insufficient to be absolutely certain that the universe would collapse, but sufficient to be sure that if it *were* to collapse, it certainly wouldn't do so for a very long time.

Confirming the future of the universe by measuring the stuff in it

Cosmologists could have rested on their laurels and accepted the persuasive simplicity and elegance of a closed universe produced from nothing at all, which would collapse eventually but take eons to do so. But scientists are not like that.

They recognized that the fate of the universe could be predicted if only we knew the amount of matter–energy that exists in the universe. Only if there is sufficient would the total gravitational attraction be enough to make the universe collapse.

A step forward had been achieved in 1933 by Swiss astronomer Fritz Zwicky after he studied the Coma cluster of galaxies. He calculated the gravitational mass of all the galaxies in the cluster and found it was at least four hundred times greater than their luminosity would suggest. He concluded that there must be matter we can't see that accounts for the 'missing' mass. He called it 'dark matter'. Many years later, when black holes were discovered, it was thought they were the prime candidates for 'dark matter'. Black holes are the remnants of giant stars, over 50 times the mass of our sun, which burn out their supply of nuclear energy and collapse under the effects of their own gravity. So concentrated is their mass that their gravitational attraction prevents even light escaping. Hence the 'black' in black hole. And anything that comes near to a black hole would be sucked in by its gravity, to be added to its mass but disappearing without trace. Hence the 'hole'. Clearly, black holes cannot be seen since no light can escape from them, but they can be detected by the effect of their gravity on other objects and by the radiation they emit. However, it is generally agreed that a collapsing universe should have more matter–energy than can be explained by black holes. There is still no firm agreement about what this could be, nor, most importantly, whether there is enough of it to make the universe collapse. Sadly for Zwicky, his proposal of 'dark matter' was largely ignored and it was only in the 1970s that dark matter became recognized as a major factor in predicting the future of the universe.

What became accepted then was that the amount of visible matter would be insufficient to result in collapse and that at least 90 per cent of all matter in the universe could not be in the form of visible stars, but in some form of 'dark matter', matter that exerts a gravitational attraction but cannot be seen or detected by scientific instruments. And that's how things stood for the next 20 years. Though theories were produced, no one seemed able to positively say what this 'dark matter' could be. It became known as the 'missing mass' (we now know it as 'matter–energy').

Looking for the answer in the remnants of the Big Bang

At the beginning of the 1990s it occurred to some bright cosmologists that if they could somehow look at the fall-out from the Big Bang, they might be able to estimate the amount of matter blasted out from it. In Chapter 1 we saw how, in 1965, Penzias and Wilson detected the microwave background that was the 'echo' of the Big Bang. Other cosmologists wondered if they could measure this in order to estimate the amount of matter–energy produced in the Big Bang. To do this they would not only have to look back in time but be able to use what they saw to make an estimate of the amount of matter. We saw how, in 1992, the cosmic microwave background explorer (COBE) project, set up to verify Penzias and Wilson's work, using the most powerful satellite telescope to probe into deepest space, gave details of the microwave radiation patterns originating from the Big Bang. The success of the COBE satellite sparked the development of a new breed of telescope, probing ever deeper into the earliest stages of the universe. Where the COBE satellite only showed simple ripples of temperature sufficient to verify that the Big Bang occurred, the sensitivity of the new instruments was such that attempts could be made to link these earliest temperature patterns to the amount of matter–energy giving rise to them. In April 2000 the results were published of a piece of COBE-like research, using the latest instruments in a balloon at an altitude of 25 miles over the Antarctic. It was called the 'Boomerang' project – Balloon Observations of Millimetric Extragalactic Radiation and Geophysics. The researchers attempted to translate accurately the temperature variations of the early universe into patterns of matter. They had produced models of the temperature patterns of the early universe that they believed would correspond to those three types of expansion – collapsing (closed), expanding forever (open) and finely balanced between the two (flat). Their task was to compare the experimental results with these theoretical models. Their results confirmed what had been evident but not conclusive from the information available up to that point – the universe was indeed flat. But there was not enough matter produced at the Big Bang to make it collapse into a Big Crunch. A similar experiment – the Maxima project – Millimeter Anisotropy eXperiment IMaging Array – was conducted shortly afterwards over Texas. Their results confirmed those of the Boomerang project. The flatness was 'perfect', which means that the universe will expand forever, slowly cooling down, ultimately resulting in the 'heat death' of the universe. This theory was first put forward by physicist William Thompson (later Lord Kelvin) in the 19th century. It might also be called a 'big freeze'. Cosmologist Paul Davies describes it thus:

> 'The universe of the very far future would ... be an inconceivably dilute soup of photons, neutrinos and a dwindling number of electrons and positrons, all slowly moving farther and farther apart. As far as we know, no further basic physical processes would ever happen. No significant event would occur to interrupt the bleak sterility of a universe that has run its course yet still faces eternal life – perhaps eternal death would be a better description.' (1)

This is important. If the universe is absolutely flat it cannot collapse into a Big Crunch. Since a state of nothing – no mass–energy and no space – can only apply to a closed, collapsing universe, these results showed that there must have been some original matter–energy that exploded in the Big Bang from a point of microscopic size – perhaps creating even more matter–energy in the form of gravitational energy in the process. An absolutely flat universe means that the Big Bang could have been exactly as Hawking and Penrose had originally suggested, a singularity at which the laws of science break down, leaving a universe that prior to the Big Bang is unknown and unknowable, and forcing us to accept that science may never be able to explain the actual origin of the matter–energy in the universe.

The state of absolute flatness predicted by that early work and confirmed by more recent cosmic background radiation research is very different from that of the other two possibilities indicated by relativity theory. The three possibilities do not have equal probability. Cosmologist Robert Dicke has likened the probability of a universe being absolutely flat to a pencil balanced on its point (2). It's a very special state indeed. A tiny deviation either way results in either a rapid collapse or an expansion so fast that galaxies have no time to form. The theory of evolution indicates that the flatness of the universe we live in is also special in that it is probably the only one able to grant us enough time for life as we know it to evolve. This is the basis of the 'anthropic principle', viz. we observe the universe the way it is because if it were different we wouldn't be here to observe it.

The death knell for a recurrent universe and a recurrent 'you'?

This research, in suggesting that the universe is infinite, seems to mean the end of the idea of a repeat universe and a repeat of our lives. As we saw earlier, the possibility of such a repeat arose from a universe emerging from nothing in a Big Bang and destined to collapse into nothing once more in a Big Crunch. A perfectly flat universe would seem to discount this possibility. However, the microwave background result could still leave open the possibility that the universe is curved but that the curvature is so small that all conceivable measurements, like the COBE, Maxima and Boomerang research cannot detect it. Just as the earth looks flat to a tiny observer, so the universe could be so big that it looks flat. In that case the possibility, indeed certainty, of a repeat of the universe and our lives within a circular dimension of time would remain.

But, in 1998, ground-breaking research appeared to discount even this possibility – convincingly.

CHAPTER SEVEN

Supernovae send a game-changing message from the past

'The future is not what it used to be.' ***Peter 'Yogi' Berra***

A quite different line of research began in 1998 and produced such remarkable results that it is generally accepted that it has taken over from cosmic background radiation research as the predictor of the future of the universe. It involves observations of a particular type of supernova using the most powerful telescopes available. Recall from Chapter 1 that the expansion of the universe is shown by the 'red shift' of distant stars. Because distance also tells us how old these stars are, the rate of expansion could be established if the red shift of distant stars could be compared with the red shift of even more distant stars. The problem is that the distance of a star must be estimated from its brightness. This presents two difficulties. First, it is hard to see stars very far away because they are so faint and, secondly, it is difficult to be absolutely sure about distance anyway. Is a star we are looking at faint because it is far away, or faint because it isn't very bright to begin with? We obviously need to look at stars that a) are bright enough to be seen at huge distances and b) have the same – and known – level of brightness. The brightest stars are supernovae, more than ten times the size of our sun, that explode in great fireballs at the end of their life when most of their nuclear energy is spent. It has been understood that supernovae of one particular type, so-called 'Type Ia', all appear to explode with the same power and, as a result, with the same brightness. Their brightness as seen from earth, therefore, depends only on their distance from us. They are 'standard candles' offering a means of accurately determining the way the expansion of the universe has changed over time. In 1998, two teams of researchers looking at Type Ia supernovae at the limits of those telescopes, found something that would turn cosmology on its head. What they found was that the red shift of the distant supernovae was less than expected and that of nearer supernovae was greater, indicating that the expansion of the universe is not slowing down, but speeding up. If this is so the universe will never collapse; it will expand forever. The shock for the researchers was so great that some of them rushed off to see if the results were caused by exotic grey dust making the distant supernova seem dimmer – it wasn't. They then looked to see if the earlier supernovae were different back then – they weren't. The results seemed solid.

Science magazine called the supernova cosmology project the 'Breakthrough of the Year'. As a result, cosmologists raced off to scour the heavens for more of these early supernovae to check those results. And so far all their research has confirmed those early findings. Three members of the teams received the 2011 Nobel Prize for physics, including Smoot. The universe was definitely expanding. It was an infinite, open universe.

Dark energy explains the acceleration and solves the 'missing mass' problem

The results were swiftly seized upon as offering a solution to the problem of the 'missing matter–energy'. We saw in Chapter 6 how Fritz Zwicky discovered that there wasn't enough visible matter in a group of galaxies to account for their gravitational effects. The missing mass became known as 'dark matter'. It was subsequently realized that this problem pervaded the whole universe – there was simply not enough visible matter in the universe to account for its gravitational effects. The problem remained a thorn in the side of cosmologists for years. How could 90 per cent of the universe go missing?

The supernovae results seemed to solve their problem at a stroke. For the expansion of the universe to accelerate there must be some hidden repulsive force. It was quickly dubbed 'dark energy'. After a few trips back to the cosmological drawing board it was concluded that the universe was made up of 10 per cent ordinary matter, 20 per cent dark matter and the rest, 70 per cent, was this repulsive 'dark energy'. Subsequently these proportions were adjusted to 5 per cent ordinary matter, 27 per cent dark matter and the rest dark energy, and this composition of the universe has become the accepted wisdom, even though almost nothing is known about dark matter or dark energy. It is hoped that the work of the Large Hadron Collider at CERN will reveal at least something about dark matter, but the identity of dark energy remains a complete mystery. It is assumed to be a property of space so that as the universe expands and more space is created, more dark energy is created, so increasing the expansion rate. The most popular theory is that Einstein's 'blunder' – the cosmological constant he inserted into his equations – was not a blunder at all but a brilliant prediction of accelerated expansion. Whatever the answer, most cosmologists have been more than happy to accept an open, accelerating universe and the existence of dark energy propelling it.

Poincaré recurrence

The Big Crunch, closed universe seemed dead in the water, and along with it, surely, was any possibility of a repeat of our lives in a circular dimension of time. A rapidly expanding universe would ultimately lead to a long and lingering thinning out of the contents of the universe – its 'heat death'.

Yet there was a glimmer of hope for the notion of a repeated universe, and ironically it came from Paul Davies, who had initially been reluctant to believe the idea of a closed, Big Crunch universe. He, along with a few colleagues, had welcomed the supernovae results, confirming what they had suspected all along – that the universe is infinite and will expand forever. Interestingly, four years earlier, discussing the implications of an infinite universe, he had acknowledged that:

> 'Infinity is qualitatively different from something that is merely stupendously, unimaginably huge … For (the universe) to endure for all eternity means that it would have an infinite lifetime. If this were the case, any physical process, however slow or improbable, would have to happen sometime, just as a

> monkey forever tinkering on a typewriter would eventually type the works of Shakespeare.' (1)

But could 'any physical process' mean the history of the universe or the history of our lives? Shortly after the accelerating universe results appeared, Davies made the comment quoted in the introduction:

> 'If the universe is infinite in spatial extent and uniform, then it is absolutely certain that there will exist other beings identical to you and me. It is one hundred percent certain that the entire inhabitants of the earth will be repeated, with duplicate Paul Davies's…' (2)

Davies was not alone in seeing the implications of an infinite universe for humanity. Swedish cosmologist Max Tegmark threw his hat into the ring, drawing a similar conclusion. In a Scientific American article in 2003 he said:

> 'Is there a copy of you reading this article? A person who is not you but who lives on a planet called Earth, with misty mountains, fertile fields and sprawling cities, in a solar system with eight other planets? The life of this person has been identical to yours in every respect.
>
> The idea of such an alter ego seems strange and implausible, but it looks as if we just have to live with it, because it is supported by astronomical observations. The simplest and most popular cosmological model today predicts that you have a twin in a galaxy about 10^{28} metres from here. This distance is so large

> that it is beyond astronomical, but that does not make your doppelgänger any less real.' (3)

Here Davies and Tegmark were referring to infinite space. Would their predictions apply to infinite time? Does a never-ending, infinite universe mean that everything that has happened in the universe so far must happen again? Including us? Referring to his comment about duplicates in an infinite universe, in 2005 I asked Paul Davies the question: if our universe is infinite in time as well as space, does that also mean that duplicates of ourselves must also occur in the future? He replied:

> 'Correct. The cosmological circumstances must ensure that there is so-called Poincaré recurrence. Current evidence ... suggests that assumption is correct.' (4)

So, just as the accelerated expansion seemed to be removing the possibility of a repeated existence, these cosmologists were proposing it as a *consequence* of the accelerated expansion.

Recall that Poincaré recurrence refers to the rule that applies to the typing monkey. It comes about when the elements of a process – like the letters produced by the typing monkey or the balls rattling round in the National Lottery machine – can continue to be re-arranged until, given enough time, eventually any pattern possible, however unlikely, must recur, like Shakespeare's sonnets produced by the typing monkeys or the same set of lottery numbers recurring in the same order a million times in a row. If, as Davies believes, the atoms, photons and other stuff that make up the universe follow the same rules – if they are subject to Poincaré recurrence – then, given enough time, an identical universe to the one we now live in must recur by pure chance. And we must be in it. Canadian theoretical physicist Don Page has embraced the notion of a repeated universe, by actually calculating when a duplicate of our universe – and therefore ourselves – will recur if time is infinite (5). It's not just a long, long time. It's the longest time that has ever been calculated or specifically imagined. The number Page has arrived at for the time it will take for the universe to recur is about:

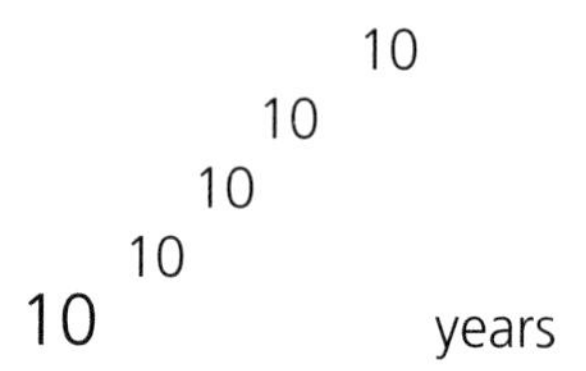

$$10^{10^{10^{10^{10}}}} \text{ years}$$

It's written this way because it would waste too many pages writing down all the zeros involved. Even if you are not a mathematician you will probably appreciate that this is one very long time. If you wanted to write it out in the 'normal' way it would have a 1 with 10,000 zeros after it – typed out it would be 20 metres long. The longest time even cosmologists normally think of is the age of the universe – some 13,000,000,000 years. Compared with the number above, the age of the universe is but a blink of an eye.

But however weird this number and the timescale it represents seem, Page, like Davies, Max Tegmark and their colleagues, is simply following the path of scientific logic using maths as the essential tool.

Let's be quite clear about what's been said. A recurrence, a precise recurrence of our universe means a recurrence of the whole history of the universe in precise detail. It's obviously a repeat of the Big Bang, production of stars, planets and life, but it's also a recurrence of every single event in the whole life of the universe exactly as they have occurred in this universe, including our lives.

But here's the rub. Could mere exact duplicates of us in these recurrent lives really be us? Recall that in a Big Crunch universe from nothing we saw that our repeated selves would feel like us because in a circular dimension of time they would *be* us. Without a circular dimension of time this would not automatically follow. Would the duplicates in an infinite dimension of time, albeit that they would look exactly like us to an outside observer, really be us, and would they feel like us?

This leads us to ask the time-honoured question, what makes us *us*, and what makes us feel like *us*? This brings in the concept of consciousness, which is embroiled in controversy. Is it possible to cut though this controversy to find an answer to our question?

CHAPTER EIGHT

What exactly makes you *you*? – the meaning of 'consciousness'

'Cogito ergo sum – I think therefore I am.' ***Descartes***

The prospect of repeating your life in a future 'recurrent' universe, rather than the *same* universe in a circular time dimension, raises a crucial question: would – or even *could* – a recurrent 'you and me' actually be, in any way, the 'me' writing this and the 'you' reading it?

This can only be answered by understanding exactly what is meant by 'you' or 'me'. Consider the question: if you totally lost your memory, would you still be you? Crucial to this question is our sense of self; of the uniqueness that makes me *me* and you *you* and stops *me* being *you*. The notion of another 'me' and 'you' in a 'future' universe might seem to challenge this crucial uniqueness. So great is our sense that we are a unique self that we are likely to feel that even an identical being in an identical setting cannot be us. An explanation depends on what really constitutes the essence of your 'self'. This has been the source of endless debate, usually in terms of the nature of consciousness, and there remains an almost universal view that a 'self' is a unique entity that could not possibly be the same self, however or whenever reproduced, even in a repeat of this universe. This, of course, has been a central pillar of the main religions for centuries, so it has a powerful provenance.

The growth in the field of artificial intelligence has led to renewed interest in answering this age-old question logically with scientific rigour. There are two main schools of thought. One is that we are no more than exceedingly complex computers, a view perhaps most forcefully championed by Marvin Minsky, co-founder of the Artificial Intelligence Laboratory at the Massachusetts Institute of Technology. The other is that consciousness, our concept of 'self', can never be recreated in a computer, because our brains are much more than a computer. John Searle, Professor of Philosophy at the University of California, is a leading proponent of this view, which encompasses the religious perspective in that the uniqueness of the individual is encapsulated in the equivalent of a religious 'soul'.

Minsky asserts that the view that our minds are more than complicated computers stems from a failure to appreciate the complexity of the operations our minds perform (1). He argues that our consciousness is simply the result of the enormous complexity of the human brain. The unique characteristic of the human brain is that it tries to understand the world around it to maximize its survival chances. Other animals do this, if at all, only at a rudimentary level. Uniquely, the human brain can look at itself. Once the human brain turns its attention to itself it begins to construct a model of self, which eventually becomes the strong identity that we recognize as us. What happens, Minsky argues, is that we grossly underestimate the nature and scope of these operations, and then argue that since even the most powerful computers cannot replicate them, there must be something in our minds that cannot be reproduced by any computer. Minsky points out that the most powerful computers, for all their sophistication and capacity, have not yet come close to replicating the operations of the human mind for the simple reason that those operations are so complex that we don't fully understand them. If we don't know what they are we obviously cannot reproduce them in a computer. But that does not mean that we will *never* be able to do so, or that there is something extra – a 'soul' – that will defy any attempt to do so.

Non-religious people seem to be split between Minsky's view and Searle's philosophical view. It is certainly difficult to believe that our powerful sense of self can be explained merely in terms of the physical processes of the brain. A powerful approach to problems of this kind, favoured by philosophers like Derek Parfit (2) is to use 'thought experiments'. This approach has a distinguished provenance since it was used by Einstein to derive his concept of relativity. A simple example is to imagine you are suffering from rapidly progressive brain degeneration. You hear of a neurosurgeon who has successfully, albeit illegally, transplanted brains. Would you consider receiving a transplant of a healthy brain? Would you imagine that you would in any way continue to exist after your brain had been disposed of? Probably not. Whatever the sense of self is, it certainly resides in our brain.

Now suppose that the diagnosis had been the other way around. You have a physically wasting and ultimately fatal disease. Your body will deteriorate relentlessly but your brain is unaffected. The neurosurgeon offers to transplant your brain into the healthy body of a young man who has died of a brain tumour. You agree, but as the operation is totally unethical it is done in absolute secrecy. It is a success. There is no trace of your previous physical self. Everything, down to your voice and mannerisms, has gone.

Your changed physical appearance may be disconcerting to you, but you would recognize yourself. So you return to your family. Would they recognize you? Despite your physical appearance, they would very quickly recognize you because you would be able to communicate and otherwise interact, as you. You would be able to talk about those things, particularly the small intimate details that only you could know. This points strongly towards the essence of self residing not simply in your brain, but in your memory alone. So consider another scenario. You fully recover from your brain transplant but remain in hospital to adjust to the changes. Suddenly your body begins to reject your brain. Emergency treatment is given and again you recover, but you are left with total amnesia. You cannot remember any part of your life. Every trace of your previous self has now gone. You wander off and are picked up by the police. The surgeon decides on total secrecy. The police have a) you and b) a missing person reported. Is there any way of identifying these as one-and-the-same person? And would it be possible for you to recognize yourself?

The last question holds the key. Total, permanent amnesia of this kind, even without the change of appearance, would remove any possibility of recognizing yourself and therefore, with a total change of appearance, there would be no way anyone else could do so. Your identity is ultimately controlled by your recognition of yourself, and reinforced by others. Since this recognition is impossible without your memory, we may conclude that your memory is necessary to identify you. The question is, though your memory is necessary, is it sufficient to identify you? Do we need to be able to communicate, for example? Our family recognizes us, not by our memory but by our ability to communicate our memories to them. Yet if we are thinking about our own view of ourselves, this communication is clearly not absolutely essential. If I come out of my operation unable to communicate, others may not know that I am me, but I would, and since it is this personal view of me that is central to the issue, we can discount the ability to communicate as necessary for me to recognize myself. This is not to say that the view of others has no bearing on our view of ourselves and therefore ultimately of who we feel we are. Indeed, the view of others towards us is so important that it is possible that without it – if we grew up from birth on a desert island, for example – the very feeling of self that we are trying to explain may not even exist. However, assuming that we already have a sense of self, the fact is that our ability to recognize ourselves does not depend on communication with others.

We are therefore left with memory. But what is memory? It is certainly a record of information we have gathered throughout, and about, our lives. But it is much more than that. As Minsky makes clear, the brain operates a huge number of interactive processes, conscious and unconscious, involving the storing, accessing, manipulating and interchanging of information, whilst at the same time it is creating new processes which, with their own continuous supply of information, must in turn be stored. All this is done by the memory. Consider a simple level of brain activity – reading this page – with no memory function whatever. Reading would be impossible because there would be nowhere to store the input information as our brain searched for a word match, and there would be no word match anyway because this could not have been stored.

And this is the key point. Memory is not a department of the brain responsible for one of its functions. To all intents and purposes memory *is* the brain. Take away memory and the other processes, important though they are, are rendered almost irrelevant. With no memory, recognition would be impossible since there would be no store of information for comparison. Everything we looked at, for example, would appear to be a frightening nightmare pattern of the unknown because there would be no essence of recognition. Not even colour would be recognizable. Every time we blinked, what we saw next would be new and unrecognizable. We could not survive independently and since most of our functions are memory-dependent it is unlikely that we could survive even with support. Even if we were somehow kept alive, there would be no residue of our humanity left, no sense of the 'self' that we are trying to explain. Without memory we could have no idea who we are or even what we are. The 'we' that we are trying to explain would not exist. This is why we so readily compare our brain with a computer, because this, too, is nothing without memory. Memory is not merely part of the computer: it effectively *is* the computer.

The difference between our brain and a computer, far from being fundamental and unbridgeable, appears therefore to be merely a matter of how information can be put in, how it can be dealt with, and how it can be acted upon. We must bear in mind that the computers we compare ourselves with have been devised specifically *not* to function like the human brain. Our brains are good at high complexity 'reasoning' tasks but hopeless at high memory capacity tasks, so this is generally what computers have been developed to do. Whereas we are designed for survival – for anything that nature throws at us – computers are designed to do specific, circumscribed tasks. Currently computer programmes are tightly convergent. They are limited in the means by which they can obtain their information, in what they can do with it, and in particular, how it can be acted upon. We, on the other hand, in order to maximize our chances of survival, must obtain information from as many sources as possible, process this information in as many ways as possible and act on it in as many ways as possible. Our programmes are divergent. Hence we have numerous senses that can interact not only with each other but with the processes they supply, with a whole range of ways in which we can react, and with the complex processing system of conscious and unconscious thinking that connects the two. A distinctive feature of our sensing, processing and action systems are that they can be directed towards other people and also towards ourselves. The result is that we obtain information about ourselves, create associated processes, and interact accordingly with the sources of our information. A key process in our survival involves our interaction with other people. The information we receive from others about ourselves plays a key role in forming our view of ourselves. So important is this information that, as I suggested above, without it we may not have a sense of self to explain. There really is only one rational conclusion: that what is put uniquely into our memory, interacting with our developing brain, laying down processes that interact with each new item of information, creating new layers of processes that then interact with each other, building a store of inter-linked information, what we call 'understanding' of our world, is all there is to being you and me, because that is all that is needed to make you, *you* and me, *me*. And this is confirmed by the fact that if all this disappears then so, effectively, do we. And, as Minsky points out, we only have the feeling that there must be more because not only do we not understand, but we also simply have no concept of, the size and complexity of the memory-based processes of our brains.

> 'We are incredibly unconscious ... The reason philosophers have been unable to explain aspects of consciousness ... is that they're always looking for something simple. If you've had something that's evolved for 400 million years, is constructed by 30,000 different genes building all sorts of stuff, then the kinds of simple things that people are looking for aren't there and that's why they seem so unsolvable.' Marvin Minsky (3)

The essence of 'self' as a unique pattern

The fact that the essence of what we are is no more than a complex pattern of memory does not take away the characteristics of who we are and what we do that uniquely identify us and give us our importance in the world in which we live. Da Vinci's *Mona Lisa* could be described as nothing more than paint spread over a piece of canvas, yet very few people would feel that this adequately defines this unique and important piece of art. If we were to pour a solvent over it so that the paint ran, and then left the solvent to evaporate, everything that was there before would still be there yet the picture would have gone because the essence of the picture was not the paint but the pattern made by the paint. In the same way, our unique sense of self can be understood in terms of a pattern, a unique pattern of memories of our experiences, our learning, and particularly our experiences and learning about ourselves, from earliest childhood to the present day. It is our view of this unique pattern that gives us our sense of 'self'.

The cultural basis of 'self'

Earlier I referred to the powerful contribution that the views of others make to our sense of self and suggested that if someone grew up alone, without this feedback, they would have a very limited sense of self. Indeed, apart from an instinctive sense of self-preservation, they might possibly have very little sense of self at all. Animals who have little or no mechanism for receiving 'messages' of 'self' from other animals, particularly in the absence of language, appear to have correspondingly little or no sense of 'self'. So the effects on the development of self will depend on the influence of other individuals. The most obvious and highly researched example of this is the influence of parents or primary carers on the nature of their children and their sense of self. And this in turn will be affected by the culture in which this occurs.

A marked feature of life in the Western world in the last half of the last century has been the rise in the importance of the individual in society. The end of the war in 1945 saw a sea change in attitude towards children, who had hitherto been very clearly subordinate to adults. Perhaps because of the feeling of guilt that it was adults of that generation that had dragged the world into a conflict that killed some 60 million people, the future generation was perhaps seen as the way to a new and better world. Whatever the reason, within a decade, there was a transformation in attitudes towards children. Nowhere can this be seen more clearly than in the phenomenal change in child-rearing brought about by Benjamin Spock in the USA within three years of the end of the war. In a very short time few social structures in the Western world were untouched by the phenomenon of child-centredness. An explosion in children's literature characterized by its emphasis on the importance of the child was soon paralleled in television and cinema, all of which was soon to be eclipsed by a total youth culture centred on fashion, music and new technology. No doubt it was driven by market forces, but the market was created out of the new attitude to children. By the late 1960s, child-centredness had produced its first generation of highly self-aware and highly empowered adult individuals. The phenomenon was, perhaps inevitably after Spock, most evident in the USA. The rise in the importance of the individual within society is one of the most significant features of the recent history of Western society. Gradually and inexorably, sometimes through strength in numbers, sometimes through a strategic use of legal systems and sometimes simply through loud voices, the individual has achieved a place of special importance, an importance that has not been seen before.

The result has been a greatly enhanced sense of 'self' within these cultures. Ever-growing post-war prosperity has offered unprecedented opportunities for self-advancement; failure to achieve it or dissatisfaction with its results led to the doors of psychologists for self-analysis; the publishing world found a new market for literature on self-awareness; educationists discovered the importance of self-esteem; everybody began exploring different forms of self-expression through which they might achieve true self-fulfilment. And, of course, threats to the individual were to be dealt with by self-defence. The Western world became truly self-centred.

It was inevitable, therefore, that in Western cultures there would develop not merely a different, but, more importantly, a much stronger sense of 'self' and that there would be a correspondingly greater need to explain it. What this shows once more is that the sense of 'self' is not in-built, but acquired. It is acquired by the process of learning; it is memory-based, and is greatly conditioned by the predominant culture.

The message is clear: what we are trying to explain in terms of something special that is an intrinsic part of us from birth is entirely explicable as no more than a body of learning about ourselves acquired partly by ourselves but mainly from the views of others, beginning at our birth and ceasing only when we die. When much is put in, as it is typically in a self-centred culture, we have a strong sense of self; when little is put in, we have little sense of self. There is no immutable 'self'. It is ever-changing, ever developing.

If, after all this, we are still not convinced that the

essence of 'self' – our consciousness - is merely our unique pattern of memory, with its vital thread of continuity from childhood, we have to come back to the fundamental question, what would be left of our humanity – of the essential 'we' – if our memory were to be totally wiped out? The answer we are faced with is that the very essence of 'you' and 'me' will have disappeared completely.

We saw in Chapter 5 that in a closed, collapsing universe the dimension of time would be circular, so that when we were reborn in the 'next' cycle we would not be a mere identical copy but the real 'we' and would feel so, just as we do now. But in an accelerating, infinite universe, Poincaré recurrence would only guarantee identical *copies* made up of rearranged matter–energy. This chapter shows that so long as those copies also have identical memories to ours, as they would in an identical universe, then however strange it seems, those 'copies' would feel like 'us', be treated as 'us' by others and to all intents and purposes *be* us.

So whatever kind of universe we inhabit, if our lives are repeated in precise detail, we can be sure that the repeated lives will not be duplicates of merely theoretical significance. They will actually be, and feel like, us – just as we do now.

CHAPTER NINE

Holes appear in the infinitely accelerating universe theory

'We have discovered that all the signs suggest a universe that could and plausibly did, arise from a deeper nothing – involving the absence of space itself – and which may one day return to nothing...' ***Lawrence Strauss***

The theory of the nature and the future of the universe derived from the supernovae research, indicating that the expansion of the universe is accelerating and not slowing down, has had a very good innings. Its provenance seems unimpeachable – the researchers discovering the acceleration are Nobel Laureates. It also seems to explain the problem of the 'missing matter–energy'. But problems are now arising.

The first signs of trouble

The microwave background results showing a flat universe were always a problem for the accelerating theory. A flat universe, as we saw earlier, is infinite but always on the brink of collapse. So an accelerating universe is not flat. This anomaly seems not to have gained much attention. But subsequent work has.

Quite independently, Steinhardt and Turok (1) Penrose (2) and Kaloper and Padilla (3) have offered different theories for the way in which the universe may not, in fact, expand forever – we may call them the STPKP theories. Steinhardt and Turok's proposal is that our universe expands sufficiently to collide with a neighbouring universe to bring about a big crunch from which a subsequent new Big Bang emerges to begin the expansion all over again and that this process continues ad infinitum. The more recent proposal by Roger Penrose is that the universe expands until all the matter decays and is turned to light – so there is nothing in the universe that has any time or distance scale associated with it. This permits it to become identical with the Big Bang, so starting the next cycle. Kaloper and Padilla accept a currently accelerating universe but argue that it is part of a natural evolution that began at the Big Bang and will end in a Big Crunch. Dark energy, they say, far from being proof of continued accelerating expansion, is, in fact, an indicator of an 'impending' process of collapse. ('Impending' in universal terms means a mere few billion years.)

Different though these theories are, they all argue that an accelerating universe would not expand forever. It would eventually return to a Big Crunch making the universe flat but closed. And, apart from the more complex process of expansion and perhaps collapse, to all intents and purposes it will be identical to the simple closed universe model described in Chapter 4, with the Big Crunch and Big Bang one and the same point of zero size and zero matter-energy in a circular dimension of time.

More problems – different types of supernovae?

But can we have full confidence in the evidence for accelerating expansion anyway? The validity of the supernovae results depended on their crucial property of representing standardized levels of brightness – 'standard candles'.

University of Arizona astronomer Peter A. Milne has recently discovered another class of Type Ia supernovae of a different colour. The whole point about Type Ia supernovae is that they are all the same. Milne's discovery undermines the crucial idea that these supernovae can be used as standard candles of brightness. It means, Milne has pointed out, that some of the predicted acceleration of the universe can be explained by colour differences between them, leaving less acceleration than initially reported. This would, in turn, require less dark energy than currently assumed (4). This immediately puts the accelerating universe theory and dark energy on a less sure footing. It also suggests that more supernovae could be discovered with different levels of brightness, further undermining the accelerating universe theory – and thereby breathing new life into the closed, collapsing universe theory.

But there's an even more devastating attack on the theory of an infinite, accelerating universe.

Is the acceleration of the universe an illusion?

Alexander Kashlinsky and his colleagues at NASA's National Observational Laboratory have shown that the huge region of space–time in which we live – a region at least 2.5 billion light-years across – is moving relative to the rest of the universe in a phenomenon now called 'dark flow'.

Greek astrophysicist Christos Tsagas of the Aristotle University of Thessaloniki has taken up this discovery and worked out how this shows that the accelerating universe is an illusion (5). He argues that it is likely that the apparent accelerated expansion of the universe is caused by the relative motion of us to the rest of the universe. The paper cites data showing that the huge region of space around us is moving relative to everything around it. The perceived acceleration was assumed to be caused by us, the observers in the present, accelerating away more slowly from those very distant supernovae than we are from less distant galaxies. But exactly the same measurements could have been caused by our localized region of space moving relative to the rest of the universe. Our region of the universe could appear to be accelerating whilst the rest of the universe is not doing so. Tsagas's theory suggests a systemic failure of measurement, not a failure of accuracy. If the results were affected this way they would *all* be affected. And so would all further results. You could not increase certainty with more confirmatory results. They would *all* be wrong, like readings from a micrometer not zeroed before use. If Tsagas is correct, the universe beyond our region isn't accelerating outward at all; rather, it is actually, in his words, 'safely rolling to a stop' (6), before starting its return to a Big Crunch.

His work has put the closed, collapsing universe theory right back into play.

Tsagas' findings are supported by astrophysicist Janna Levin who has also advocated the theory that the universe is not infinite but finite and closed (she uses the word 'compact') like the earth, with no edges, adding that as a result, one of the 'distant' galaxies we detect may in fact be our own galaxy seen across circular space–time, though strictly, it would actually be our own galaxy billions of years ago, because of the time it would take for light to make the journey (7).

Lawrence Strauss of Arizona State University has also firmly staked his closed-universe-from-nothing flag to the mast. Strauss says:

> '...quantum gravity [the quantum theory of gravity] not only appears to allow universes to be created from nothing ... meaning the absence of space and time ... it may require them [to be so] ... Moreover, the general characteristics of such a universe, if it lasts a long time, would be expected to be those we observe in our universe today.' (8)

Certainly many scientists are troubled by the need to invent totally mysterious dark energy, the nature of which we have no idea, to explain the perceived acceleration of the universe even though it seems to solve the long-standing problem of 'missing' matter–energy in the universe.

There are two more phenomena evident in nature, explored in the next two chapters, that count against an infinite, accelerating universe, and support a closed, collapsing universe recurring within a circular dimension of time. The first, which we shall now look at, is the little-recognized fact that circularity, not linearity, permeates every aspect of the physical world.

CHAPTER TEN

The universality of circularity

'My neighbour has a circular driveway – he can't get out.'
Steven Wright

If we look around us with the insight of modern physics, we find that it is circularity, not linearity, that dominates the physical world, and that a circular dimension of time does not distort the picture but completes it.

The evidence of the astronomical and atomic scales

On both the largest and smallest scales, from the astronomical to the atomic, circularity holds sway. On an astronomical scale we have swirling galaxies, spherical stars, spherical planets and circular planetary orbits. On an atomic scale we have spherical electron clouds creating spherical atoms. Only in the narrow scale band in which we exist do we see linearity apparently predominating. From our human perspective, the earth looks flat and the building blocks of our structures appear rectilinear. Our view of nature therefore depends on our choice of scale.

The evidence of relativity

We saw in Chapter 2 how Russian physicist and mathematician Alexander Friedmann showed that one of the predictions of Einstein's general theory of relativity was that gravity is so strong that space is bent around on itself into a circle, so that if we travelled around the universe in a perfectly straight line, we would end up where we started. Space itself, in such a universe, would be circular. There could be no such thing as 'straight' lines. The very fabric of the universe would be circular. Linearity would be merely an approximation that would work reasonably well on the smallest scale. So relativity has built into it a propensity towards circularity.

The evidence of symmetry and the fundamental laws

When we consider the three dimensions of space (height, breadth and depth) we see that there is no 'preferred' dimension, no dimension that is somehow different or more important than another. Space is fundamentally symmetrical. Circularity represents the ultimate in symmetry and a sphere is the most symmetrical figure.

Circularity rules through symmetry because the fundamental laws are based on circularity. As we have seen, gravity rules the cosmos. The law of universal gravitational attraction is a circular law. The gravitational attraction experienced near an object depends, spatially, only on the distance from the object. Direction is irrelevant. Equidistant points lie on a circle or, strictly, on the surface of a sphere. Gravitational attraction decreases in proportion to the increase in the surface area of the sphere on which the object we are considering lies. So it is inversely proportional to the square of the radius of the sphere. Gravity's effects are indisputably circular. This is why planets are the shape they are. As they were formed, their gravity pulled them into the most compact, lowest energy state – a sphere. The circularity of gravity lies at the heart of the circularity of the dimension of space predicted by relativity.

The other fundamental laws, the best known of which are the laws of electrical and magnetic attraction, are exactly the same 'inverse square' laws as gravity, based on circularity. Just as gravity governs the large-scale universe, so these laws govern the small-scale universe.

The evidence of periodicity – clocks, vibrations and waves

Perhaps the most obvious evidence for a fundamental circularity can be seen in the way that, even if we think of time as linear, our methods of measuring it have been fundamentally circular. The measurement of time over the whole of our existence has been based on the earth revolving – around the sun and on its own axis. And, of course, it is the moon revolving in its circular path around the earth that has given us our months. When we needed more sensitive and accurate measurements we looked first at swinging clock pendulums, then at oscillating balance wheels in watches, and now at oscillating atoms in quartz chronometers and in atomic clocks in physics labs. Circularity lies at the heart of all of these. A weight swinging on the end of a pendulum, a balance wheel reciprocating against a circular spring, and the oscillations of a quartz or caesium atom – are all examples of a special form of periodic motion known as 'simple harmonic motion'.

Simple harmonic motion is the most basic of periodic motions and it is based on circularity. It is defined as the

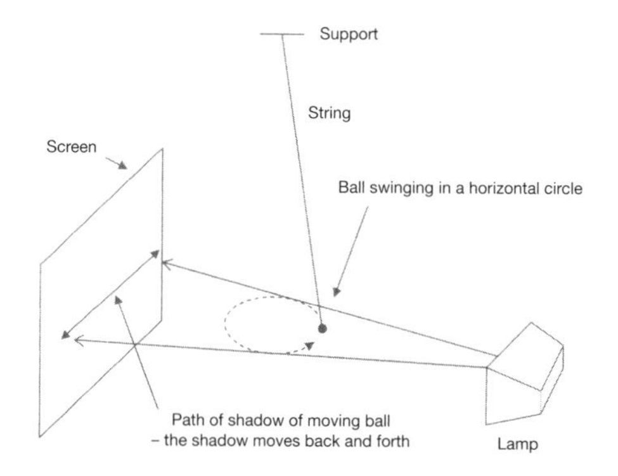

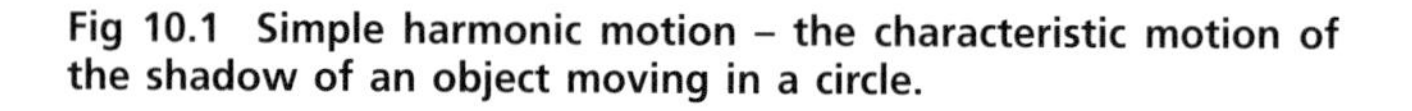
Fig 10.1 Simple harmonic motion – the characteristic motion of the shadow of an object moving in a circle.

'projection on a diameter, of uniform motion in a circle'. If we imagine an object moving at uniform speed in a circle – a ball at the end of a long string, for example – and project a shadow of its movement not onto its diameter but the equivalent, onto a screen, the shadow will move in a straight line backwards and forwards (Fig 6.1). The motion of the shadow is simple harmonic motion. If we plot a graph of this motion against time, we get a wave, the characteristic wave of simple harmonic motion. It is often called a sine wave because the vertical displacement of the wave depends on the sine of the angle of the circular motion.

All periodic behaviours in nature – oscillations, vibrations and waves – are based on this simple fundamental harmonic 'sine' wave which, as Fig 6.2 illustrates, is firmly rooted in circularity. Such periodic behaviour and the waves associated with it are fundamental in nature.

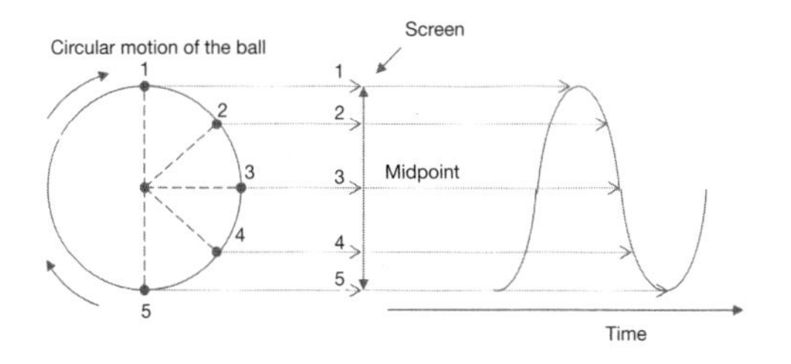

Fig 10.2 Circular motion projected onto a screen and thence translated into a displacement – time graph giving the characteristic sine wave of simple harmonic motion and showing that this motion is rooted in circularity.

A universe of waves and vibrations

Nature abhors a vacuum. On the other hand, she adores waves and vibrations. They are to be found at the heart of almost every aspect of nature. We are familiar with the waves on our oceans, but the natural world is much more fundamentally wave-like than that.

Most of the information we gather about the world is obtained by our two most important senses, sight and hearing, which are designed to detect wave phenomena. Sound, of course, is the vibration of air. Light is electromagnetic energy that is transmitted in waves. Not only light, however, but all forms of electromagnetic radiation – radio waves, microwaves, radar waves, infrared, ultraviolet, X-rays – are waves of energy. Everything that happens – the whole history of our universe – is based on the transfer of heat energy. Heat energy is to be found in two forms; heat stored in matter and heat radiation. Heat in matter is stored as vibrating molecules. Every particle of everything that has a temperature above absolute zero on the Kelvin scale (–273°C) is vibrating. Heat radiation is, of course, infrared radiation – waves of electromagnetic radiation in the infrared wavelength. Waves of infrared radiation are emitted by hot objects – molecules vibrating. When any form of electromagnetic radiation is absorbed by matter, its wave energy is turned into heat, the vibration of molecules. Waves or vibrations – we simply cannot get away from periodic behaviour in nature. And it is all based on circularity.

Quantum theory and waves of matter

Appendix B explains that in quantum theory, waves even lie at the very heart of matter because *matter itself has wave properties*. It does not simply vibrate, it does not simply emit electromagnetic waves of energy, it actually *is*, itself, a wave phenomenon. Wave motion, and therefore circularity, cannot get much more fundamental than this. Or can it? Enter string theory.

String theory and vibrations

String theory forms part of M-theory, the latest and most promising attempt at a 'Theory of Everything'. The essence of string theory is that it proposes that the fundamental building blocks of matter are not particles but vibrating strings. The particles we currently regard as 'fundamental' are therefore not fundamental at all. They are made up of tiny, one-dimensional loops that vibrate in different ways, giving rise to the different particles that we had previously thought of as fundamental. The mass and other characteristics of these particles are determined purely by the patterns of vibration of their strings. In string theory, particle properties are no more than the manifestation of a single physical feature – the resonant patterns of string vibration.

Since vibration, as we saw above, has its roots in circularity, string theory means that circularity is even fundamental to the fundamental. Not only does matter have a fundamental wave nature, but the building blocks of matter itself may be no more than vibrations. Since all matter in the universe originated from the 'condensation' of radiation energy – wave energy – perhaps this should not be totally unexpected.

String theory ventures even further into the realms of circularity. It proposes that the dimension of scale – from the smallest to the largest, and apparently quintessentially linear – is in fact circular. Bear with me as I introduce a little maths...

The theory proposes that a universe of radius R × the Planck length (where R is any multiple) is physically identical to a universe of radius 1/R × the 'Planck length' (this is the smallest measurement possible in quantum physics). So if $R = 0$, $1/R = \infty$ - making the universe at zero size physically identical to the universe at infinite size. This makes the dimension of scale circular. (This phenomenon and its possible ramifications are discussed in more detail in Appendix A.)

So whatever natural phenomena we look at, be it vibrations or waves, periodicity and, therefore circularity, lie at their heart.

In isolation, the examples of circularity above are merely suggestive. Together, however, with no contradictory examples of fundamental linearity, they represent powerful counter-evidence to the notion of a linear dimension of time. Again, linearity appears so prevalent and therefore so basic to us only because of our extraordinarily narrow perspective within a huge universe. As we take a wider perspective, its prevalence diminishes whilst that of circularity increases until eventually no linearity remains, only circularity. It is perhaps therefore more relevant to ask, not how could time be circular, but how could it be otherwise?

But that's not all. An ever-accelerating universe expanding to who-knows-where, driven by the totally mysterious 'dark energy', of which who-knows-what, flies in the face of yet another principle describing the large-scale features of the universe – the principle of simplicity.

CHAPTER ELEVEN

Science and simplicity

'According to modern astronomers space is finite. This is a very comforting thought, particularly for people who can never remember where they have left things.' ***Woody Allen***

One of the criticisms levelled at Hawking's theory of time was that he made simplifying assumptions and, bearing in mind the complexity of the universe, the resulting model itself seemed improbably simple. After all, the universe is an unimaginably complex network of reactions and inter-reactions governed by a huge number of physical laws, many of which we still do not understand. How could the underlying nature of such a complex system possibly be simple? It is indeed true that physics at an atomic level is complex. It is impossible to predict the behaviour of just a handful of molecules of the air you are now breathing. Quantum theory makes things even more complex at a sub-atomic scale (discussed in more detail in Appendix B). Therefore, reflecting on just how many molecules and sub–atomic particles there must be in the whole universe, it would be understandable if people felt that nature is most complex at the cosmological level. However, we would be wrong. That's not how nature works. And scientists recognize it:

> '...the behaviour of the universe on a very large scale seems to be simple.'
>
> Stephen Hawking (1)

> '...most particle physicists, including myself, believe that the correct theory of nature will be simple and elegant...'
>
> Alan Guth (2)

> '...it remains a great and fundamental truth that the universe seems to run according to rules which are so simple that the theorists initially felt that they must be inadequate descriptions of reality.'
>
> John Gribbin (3)

Richard Panek describes how, three days after receiving the Nobel Prize for Physics for his work on the cosmic background radiation that confirmed Big Bang cosmology, George Smoot was heard to exclaim:

> 'Time and time again the universe has turned out to be really simple.' (4).

And, of course, you will have read the poem at the beginning of this book by a renowned cosmologist, the late John Wheeler.

These are the views of seasoned scientists, cosmologists who have devoted their lives to understanding the nature of our universe. The rules of science generally become simpler as the scale we observe becomes larger. This is why the paths of planets and comets can be accurately predicted into the future. This suggests that at the cosmological level we can expect to find them at their most simple. Recall perhaps the most fundamental relationship known in physics, Einstein's equation showing the relationship between matter and energy:

$$E = mc^2$$

This is the epitome of simplicity. If nature were fundamentally complex this would surely have turned out to be one of those diabolically endless equations that always seem to appear on a blackboard in photographs of scientists at work. In fact, complexities are only to be found in nature on the smallest scale. The more globally we observe nature and the wider our perspective, the simpler and more elegant it gets. Stephen Hawking would have difficulty predicting the behaviour of three gas molecules over the next ten seconds, but he would doubtless be able to predict the behaviour of three galaxies over the next ten years. We must beware, therefore, of criticizing cosmological theories on the grounds of over-simplicity. Indeed, we should be suspicious of theories that are complex. The conclusions set out in this book may be wrong but it is unlikely to be because they are too simple.

Ockham's Razor

The principle of fundamental simplicity is not new. Seven hundred years ago, the English philosopher William of Ockham (or Occam) in Surrey devised a 'tool' for eliminating unnecessary complexity in philosophical debate. He was frustrated by the tangle of medieval metaphysical theories and pseudo explanations that obscured the fundamentals of Christianity, and proposed a principle that has come to be known as Ockham's Razor. He said that we should not assume more causes for any phenomenon than are absolutely necessary to explain it. Okham's principle has become embedded in scientific method. Alongside our expectation of simplicity and elegance in fundamental physics, it undermines the case for the accelerating universe and its complex ramifications.

Therefore, as we move towards a 'Theory of Everything' we should not expect scientific knowledge to be so complex as to make it inaccessible. Stephen Hawking's aim was not only to determine the underlying laws of science, but also to demystify them. Recall his aims for the Theory of Everything :

> '...if we do discover a complete theory, it should in time be understood in broad principle by everyone, not just a few scientists.' (5)

So, in the principle of simplicity we have further support for the simplest universe model. It bodes ill for the accelerating universe theory and the complex conclusions drawn from it.

The accelerating universe model was eagerly seized on by much of the cosmological community because it seemed to offer a solution to the long-standing and frustrating 'missing matter–energy' problem presented by the cosmic microwave research. Yet we know nothing at all about the 'repulsive' dark energy that has been acclaimed as the solution. The fact that it has effectively been dreamed up merely to 'explain' acceleration and the 'missing' matter–energy, with no indication of its nature, should fill scientists with anxiety rather than euphoria, especially as it appears to confound the understood laws of both energy conservation and gravity. As 'missing mass–energy' it should be contributing to gravitational attraction, not repulsion. The current favourite supportive theory for dark energy is that it represents the cosmological constant that Einstein thought was a mistake. Another is that the mysterious anti-gravity force is a form of vacuum energy or 'quintessence', a name reminiscent of the substance from which the heavens were thought to be made in order to distinguish them from the four 'earth, air, fire and water' elements that made up the world in ancient philosophy.

As the cause of accelerating expansion, dark energy appears to be an ever-increasing source of energy apparently coming from within the ever-increasing space it creates. This should raise eyebrows regarding the law of conservation of energy.

So an ever-accelerating universe adds unknowns to the picture instead of removing them.

The most important function of science is to explain and simplify, to bring superficially unconnected pieces of evidence together to make a theory that predicts, and simplifies. The supernovae research doesn't do that. It complicates. Or rather it further complicates. As we have seen, the complication began in the 1930s when Zwicky found discrepancies between the mass of large astronomical objects determined from their gravitational effects and the mass calculated from the 'luminous matter' they contain – stars, gas and dust. So those ever-resourceful astrophysicists deemed that there must be some different, invisible matter responsible and it became known as dark matter. We saw Zwicky's contribution to this problem. Yet apart from black holes, which are thought to be too scarce to solve the problem, no dark matter has ever been detected directly. But that was only the start of the problem. Cosmologists could still only make this mysterious dark matter account for 25 per cent of the total matter–energy of the universe. So, undaunted, they deemed that the remaining 70 per cent of the perceived missing mass must be something else even more mysterious. So, 95 per cent of the matter–energy of the universe consists of something we know little or nothing about. What is it? Where does this energy come from? How is the conservation law of energy preserved? Space is undoubtedly something, rather than nothing, but why is dark energy repulsive rather than attractive? If it represents missing matter–energy, shouldn't it be attractive, like other matter–energy? Other than an ad hoc explanation of accelerated expansion, and a convenient candidate for the rest of the 'missing matter–energy', what evidence is there for the existence and nature of dark energy? It seems to solve those two important problems rather in the way that the phlogiston theory 'solved' the problem of burning and the aether theory 'solved' the problem of the motion of light. Both were based on shaky science and were subsequently rejected.

CHAPTER TWELVE

Summary and conclusions: the implications of a circular dimension of time

'If there is such a thing as reincarnation, knowing my luck I'd come back as me.** **Rodney Trotter

There is no doubt that our universe is expanding, and there is abundant evidence that it began in a primeval explosion of matter–energy – the Big Bang. There are two, and only two, possibilities for its future. The first is that in billions of years, under the influence of the gravity of all the matter–energy produced in the Big Bang, the expansion will reverse and slowly but surely collapse into a Big Crunch. The second possibility and current front-runner amongst cosmologists, is that the universe is infinite – it will go on expanding forever, whether by the accelerating process currently believed or by some other process as yet unknown.

In the first scenario, in which the universe collapses into a Big Crunch, it has been shown that all the matter–energy produced in the Big Bang could have been produced from an absolute nothing, because the positive matter–energy produced is exactly equal to the negative gravitational energy of the matter–energy created and blown apart. The positive matter–energy has been created from an equal 'debt' of negative gravitational energy just as money can be 'created' by borrowing, and so creates an equal amount of debt. Not only at the Big Bang but throughout its subsequent life – expansion and collapse – the total energy of the universe – positive matter–energy plus negative gravitational energy – remains zero. When the universe collapses into a Big Crunch, it will disappear to absolute nothing once more. As 'absolute nothings' – no matter–energy nor space – the Big Bang and Big Crunch will be identical in every conceivable way. Following the Principle of Identity, they must be regarded as one and the same, so that at the Big Crunch the universe will begin again as the Big Bang. This means that the dimension of time must be circular; that the universe will recur within this circle of time and our lives along with it. Because we can have no concept of time when we are dead, the instant we die we shall awake in the 'next', but actually the same, universe. The fact of a circular dimension of time ensures that the next 'we' will feel exactly the same as this 'we' because they *are* the same, not merely identical. They will have exactly the same memories, which are necessary for us to be sure they are the same.

The supernovae results brought the second scenario to the fore: an infinite universe expanding forever, ever-faster, along a linear dimension of time. It challenges, and appears to disprove, the theory of a closed universe and a circular dimension of time.

But now we have seen *this* theory significantly, and perhaps fatally, undermined.

First, Milne showed that all supernovae may not be the same. Remember that the accelerating expansion theory depends on the type 1A supernovae having the same brightness. Milne's research shows that this might not be so. There are at least two colours of type 1A supernovae.

We then saw how the three different STPKP theories showed that even a currently accelerating universe would eventually still collapse again into a Big Crunch, emerging in a Big Bang exactly as it did in the first, closed universe scenario.

And, lastly, we saw the Tsagas results showing that the perceived acceleration could be the result of our localized region of space moving relative to the rest. The universe is not expanding ever-faster, but 'safely rolling to a stop', from which point it will collapse to a Big Crunch/Big Bang. If the results are confirmed, and the supernovae evidence is overturned, then the accelerating universe theory is dead in the water. The universe will be closed after all. The theory of a closed universe from nothing is revived and with it a circular dimension of time and a precise repeat of our universe and our lives.

We saw significant and highly persuasive support for this theory from a quite different direction. The physical world we live in is based on circularity. Apart from the obvious evidence of spherical planets and stars in circular orbits, we saw the fundamental circularity of energy and matter in their wave properties and vibrations. There are no equivalent examples of fundamental linearity.

And if that were not enough, we saw how the accelerating universe into an unknown oblivion cut across another fundamental aspect of the physical world:

simplicity.

But that's still not all. Even if, despite all this, the universe *were* to expand infinitely along an apparently linear dimension of time, there would be sufficient time for Poincaré recurrence to produce identical universe after identical universe, effectively morphing the apparently linear dimension of time into a circular dimension. Circularity would still prevail.

Which takes us to our final conclusion. *Whatever the current behaviour of the universe, it appears that it is destined to have the same future as the simplest closed universe from nothing: it will be repeated in precise detail, along with our lives, within a circular dimension of time.*

In short, when we die, we shall indeed be instantly reborn again with precisely the same lives ahead of us.

The implications

This takes us to where I believe Hawking himself wanted to go. Science has stepped in where scientists have traditionally feared to tread. And this is exactly as it should be. If both science and religion claim to explain our world then both should have legitimate views on the areas they seek to explain.

The prospect of beginning our existence again the instant we die has many implications. The most important point to bear in mind is that because we are talking about a repeated existence, not a 'future' existence, this model excludes any possibility of reincarnation, *déjà vu* or other similar phenomenon. The idea that *déjà vu* is a memory of a previous cycle of existence seems attractive but it is not on offer because there can be no memory of a 'previous' cycle. *Déjà vu* is probably no more than a false 'recognition' trigger in our brains. Or, as recent research suggests, the effect of the brain testing memory.

The significance of a repeat existence is that we know exactly what it will be like, up to the point of our current existence. We cannot know more than that.

Just our lives all over again?

It is important to recognize that in this model there is no scope for doing things differently from one 'life' to the next. There can be no learning from mistakes, no making things better the next time around. The 'next' cycle will be precisely 'this' cycle over again. We are destined simply to repeat our lives, down to the finest detail, with no memory of any previous existence. This is where unscientific thoughts are likely to creep in – 'What's the point of an endless repetition of our lives?' 'Why have another life if you can't improve on this one?' These are all understandable questions but the options they suggest are not on offer from the repeating existence theory.

So it is not a question of what we like or don't like, or can see a point to it. It is a matter of the direction in which science points. For some the prospect of reliving their lives is not attractive: the starving, the diseased, the victims of war or natural and man-made disasters. For them the prospect of an idyllic eternal afterlife in the care of a loving God would be much more attractive – and fairer. But science doesn't deal with fairness. If science precludes an idyllic afterlife then we are faced, as Galileo was three hundred years ago, with a choice between science and our beliefs. The science of the outcome proposed says no more than that the circularity of the dimension of time appears to be a certainty and it means that our death is not the end.

A predetermined universe

Classical, pre-quantum-theory science saw the universe running according to rigid laws and led to the view that everything in the universe was predetermined. It was therefore simply a matter of discovering all the laws of science to be able to predict everything. The advent of quantum theory, showing that at an atomic level it was impossible even in principle to predict anything precisely, seemed to sound the death-knell of determinism. Although, as we have seen, the universe on a large scale is eminently predictable – large-scale laws are extremely accurate – the power of the uncertainty principle of quantum theory has prevailed, and very few people today would give any credence to the classical view of a predetermined universe.

Yet the meaning of a circular dimension of time, or even a precise repeat of the universe, is quite clear. If our universe recurs, it must be a precise replay of our current universe. Everything, down to the single flap of a butterfly's wing in the Amazon jungle, and including, therefore, every aspect of our lives, must be exactly the same. The implication is inescapable. No matter what the uncertainty principle says about predictability at a quantum level, the history and future of our universe must be completely predetermined. Though we cannot know what lies ahead of us in this life – 'predetermined' does not mean 'predictable' – it must be exactly what lay ahead of us in all 'previous' cycles. It cannot be different because it is the same universe in the same cycle.

We had a hint of a predetermined universe when looking at Stephen Hawking's view of the nature of time. We saw that, just as space 'sits there with places scattered over it', time sits there with events scattered over it, waiting for us to arrive at them'. In this sense, therefore, they exist already. This is just another way of saying that everything is predetermined. In a lecture at Cambridge University in 1990, Hawking specifically discussed the question of determinism in reference to his view of time. He concluded:

> 'Is everything determined? The answer is yes, it is. But it might as well not be because we can never know what is determined.' (1)

Even if one accepts Hawking's view of time, one might argue that predetermination does not follow, and that at every point along the dimension of time quantum uncertainty offers a range of possibilities that happen by pure chance. A circular dimension of time, however, in which each and every possible quantum history of the universe exist as multiverses at the Big Bang, means that every event within each of these histories must be predetermined. Circular time and a predetermined universe are inescapably complementary.

Predetermination is also a reminder that this model does not rule out the existence of God. If everything is predetermined it is perfectly reasonable to suppose that it is predetermined by God. A circular dimension of time may require adjustments to conventional religious beliefs; it does not challenge the existence of God.

The implication for the limits of prediction

Where, then, does a predetermined universe leave quantum theory and the uncertainty principle? If the theory of circular time leads to a violation of the uncertainty principle then this must surely mean that it is the theory of time that is wrong, not the uncertainty principle. If we look carefully at the uncertainty principle, however, we see that what it says is that at the most fundamental level it is impossible to determine precisely what will happen. This does not rule out determinism. It merely sets a limit on *our* ability to determine everything, which is a very different thing. It is, of course, unsurprising that the uncertainty principle in quantum theory should be an inevitable feature of a universe that we are a part of, for if we could discover and use fundamental equations that determined precisely everything that happens in the future, this would include events that we could alter, and therefore the equations could not have been fundamental. Hawking considered this. He referred to it as 'the logical paradox of self-referencing systems'. (2)

So the fact that we are unable to discover the rules that determine the future of our universe precisely does not mean that the future of the universe is not precisely determined. As Asimov remarked:

> 'The uncertainty principle means that the universe is more complex than was thought, but not that it is irrational.' (3)

The uncertainty principle does not say that there is irrationality in the fundamental behaviour of our universe, merely that there is a fundamental uncertainty in our ability to predict this behaviour. And this is simply due to the fact that we are part of the behaviour we are trying to predict. The uncertainty principle gives us a necessary reminder of our limitations.

The implication for free will

But if the universe is pre-determined, what becomes of our free will? Most of us believe that, if not in control of our destinies, we are at least in control of ourselves; we have 'free will'. This is, of course, a central feature of most religions. The traditional explanation of the paradox of evil in a world created by a caring God is that we have been given free will to choose how we behave. However, the importance of the concept of free will is not confined to religion. Most people will feel that free will lies at the heart of our humanity. When I choose a new shirt, or the next word I write, I have no sense of any outside control over my decision. There is influence but not control. I recognize that my choice of shirt is influenced by fashion, my preference for blue, by childhood experiences and so on. But there is nothing stopping me from overriding these influences. Yet if everything I do, every choice I make, is predetermined, is free will still an illusion?

The matter is easily resolved. The problem arises only if free will is effectively *defined* as the absence of predetermination. Then, of course, the two *will* be incompatible. This is what it appears philosophers have done. If, however, we define free will more carefully, we see that there is no incompatibility. Free will is the ability to make choices from a range of possibilities *without any restriction on our choice*.

If I have shirts, words, holidays or whatever to choose from, there is no restriction on my choice. I am completely free to choose one or other or both or none. I have total free will. Even though my final choice was predetermined and inevitable because it is locked into the history of the universe, in no way could I sense that my choice was restricted. On this basis free will and determinism are not incompatible.

In *Physics for the Enquiring Mind* (4) physicist Eric Rogers points out that free will and determinism have the same relationship as waves and particles in quantum theory (see Appendix B). Though they appear to contradict each other, one does not deny the existence of the other. They are simply different perspectives of the same thing. Each is true when we are dealing with its perspective, though the other is excluded whilst we are using that perspective. This is Danish 'father of quantum theory' Niels Bohr's 'Principle of Complementarity'. Even though two perspectives contradict each other, both can be true and then, most importantly, both are necessary for total understanding. Bohr's principle is in fact the principle of 'partial understanding'. One perspective on its own provides only a partial understanding of a phenomenon. We can therefore expect to have conflicting perspectives when we do not fully understand something. We do not really know what matter actually is at an atomic level. We simply know two things about it – that it sometimes behaves like a particle and sometimes as a wave, and, as with all perspectives, what we see depends on how we look at it. In a similar way, Hawking gave us two perspectives of time: 'real' and 'imaginary'. They, too, were contradictory and created a dilemma – which one was right and which one should be used? Because we had been using 'real' time for thousands of years and we knew this served us well, most scientists simply rejected imaginary time much as they rejected the wave behaviour of matter when it was first proposed because it appeared to contradict the particle theory that they knew was true. They can be understood as complementary perspectives. 'Complementarity' means that all perspectives are valid even though they appear to contradict each other. More than that, however, all are needed for a complete understanding. This is explored in more detail in Appendix B.

And so it is with free will and determinism. They are perspectives of nature that whilst apparently contradictory, are nevertheless both true, and both are needed to fully understand the nature of our universe.

The implications for morality

An inevitable response to the suggestion of a predetermined world is that it absolves us from responsibility for our behaviour. If someone commits a crime they can argue that their actions were predetermined and that they therefore had no control over them. In fact, this is simply another aspect of the free will issue. If a predetermined universe does not affect free will because it does not remove our ability to make a free choice, then we make our choices in full knowledge of the consequences and cannot escape the responsibility for them. A predetermined universe does not force me to rob a bank. I have choice and before I make it I am aware of the likelihood of a long prison sentence. A predetermined universe would therefore not absolve me of responsibility for my crime. It merely means that an 'out-of-universe' observer, like God, would know what you were going to do.

The implications for religion

As we saw above, the notion of a predetermined world is not incompatible with religion per se; if God knows everything, including future events, those events must be predetermined. However, a circular time model does seem to pose a serious challenge to certain fundamental religious beliefs. For example, one of the cornerstones of religion is the possibility of an afterlife in heaven. But this appears to be impossible if, when we die, we are *immediately* (as it seems to us) reborn in the next cycle of time. In this restricted circle of time there seems be no afterlife in which there is some form of closer relationship with God. Could that be that it is then only a small further step to questioning the actual existence of God? Not at all.

Even if there was no creation event, there is still the question of how the whole cycle of existence came into being in the first place. And it might be possible to have an afterlife between death and rebirth. After all there are many billions of years to fit it into. Obviously, the conventional view of an afterlife, and therefore the nature of religious belief, must be revised. But the revision retains key elements of the traditional concept of religion. Consider the implications of our lives being destined to be repeated. As we have seen, even though our actions are predetermined we retain the free will to choose, and so we are still able to choose exactly what is to be repeated. We therefore have very much the same decisions to make in a circular time universe as we would with a conventional afterlife. If I now decide to embark on a life of crime, I know that I will not only go to prison in this life but in all subsequent life cycles. If I ruin my life with drugs now, I am setting the pattern of my future existences. Circular time breathes new life into the archaic concept of 'eternal damnation'. The traditional religious view is that an eternal afterlife gives my behaviour in this life a much greater significance. In a circular time model, the preservation of my free will gives my actions a very similar significance. Because I know that my mistakes are 'locked in' to my life history and are therefore destined to be repeated with each time cycle, the pressure on me to avoid them in the future is much the same as it would be with the prospect of a heavenly afterlife. Circular time may therefore have an effect on my morality that is little different from that of traditional religion. The important thing to recognize is that the fact that my actions were predetermined does not affect the choices available to me, nor the consequences of making my choice.

The implication for personal philosophy

A feature of some religions is the concept of 'God's will'. In such religions any tragedy is accepted as 'God's will' and not questioned, even if it is not understood. It is clear that there is a great comfort in this because it removes the impotent and destructive anger or guilt that such tragedies otherwise engender. Few of us will not have experienced those feelings of thinking, 'If only I had done this instead…' or, 'If only this had happened differently…' A predetermined universe means that all events are inevitable, including the tragedies, and that there is therefore no meaning to the term 'if only'. Though our free will makes us responsible for our actions *before* an event has occurred and for any of their consequences, once that event has occurred we can comfort ourselves with the knowledge that everything was destined to happen that way. There really was no possibility of an alternative – no 'if only'. A predetermined universe means that, just as we might hold fervently to the concept of 'God's will', we are able to accept philosophically that what has passed has passed and to get on with dealing with the future without the baggage of anger or guilt. It may, of course, encourage complacency. 'If everything I do is predetermined then I don't have to think about anything or agonize over decisions'. Determinism is thus a double-edged sword. Free will means that it can be used or abused.

The implication for the purpose of it all

Science does not deal in ultimate purpose. When someone asks what the purpose of life is they are presupposing that there *is* a purpose. So before the question can be answered it is reasonable to ask for evidence that such a purpose exists. If no such evidence is forthcoming then we can respond by saying that the question has no meaning, any more than asking, what is the shape of violence? Science merely tries to link together pieces of empirical evidence in order to discover underlying rules that will help us to understand – that is, to predict – what happens in other situations. Science may say that the purpose of reproduction is to maintain the survival of our species but it does not say what the purpose of our existence is. Science can go a long way towards explaining the world we live in, but it does not tell us, in Hawking's words, 'Why the universe bothers to exist at all'.

The implication for the hope of a better world

Earlier we considered the unfairness of someone whose life has been blighted by disease, famine, war or natural disaster. The iniquity of such suffering is hard enough to accept for a single lifetime; to have this repeated endlessly would, to many people, be unbearable. There would surely be irresistible pressure to reject any theory that proposed such a scenario.

However, the sheer unacceptability could have an unexpectedly positive outcome, for it might be sufficient to mobilize the conscience of humanity in a way that has hitherto not been seen. So great could be the abhorrence at the prospect of fellow human beings condemned to real eternal suffering that perhaps, just perhaps, the natural compassion of the human spirit, so often hidden in our modern selfish society, might give rise to a powerful movement to eradicate needless suffering. How remarkable it would be if the science of circular time were to be able to join forces with religion to bring about a truly better world.

Further thoughts

The field of cosmology has been rife with speculation about the implications for the human race as a whole of both accelerating and collapsing universes, with different proposals as to how human ingenuity might find ways of avoiding the inevitability of roasting in a Big Crunch or freezing in the heat death of an expanding universe. These have generally been based on a particularly optimistic view of the longevity of the human species. Most biologists, even those who insist on placing humans at the top of a 'species intelligence' scale, agree that we must be placed quite low on a 'species survival' scale. Bacteria top this scale. It is highly unlikely, even with the most far-reaching products of human ingenuity, that we could develop the powers of adaptation needed to face the changes that are likely to occur over the next million years, let alone billions of years. For the same reason, it is unlikely that any intelligent species that exist in other parts of the universe will fare any better. A circular-time universe offers the most assured way of escaping those depressing fates, being reborn as ourselves with no memory of our existence in the 'previous' cycle. It offers us eternal life – on earth.

The wholeness of the universe

We have seen that we are predisposed to see ourselves as quite separate from the rest of the universe. We like to feel that we are somehow above the laws that govern the rest of the natural world. This caused problems when we tried to come to terms with quantum theory. We have to accept that our existence as observers automatically affects what we are observing, making it impossible to know what is going on when we aren't observing it. Even so, we are still inclined to view the universe as a collection of bits of matter, forms of energy and areas of space that can somehow exist, and therefore be investigated, independently of each other, even though we know that the merest twitch of any single electron or photon has an effect on everything else in the universe. They, like the butterfly in the Amazon jungle so beloved of chaos theorists, affect the rest of the world because they are already connected to the rest of the world, not only by their electromagnetic and gravitational fields, but by space, which we now know is not nothing but something. As the naturalist John Muir remarked:

> 'Whenever we try to pick out anything by itself, we find it hitched to everything else in the universe.' (5)

Perhaps the final message that we have to accept, therefore, is that not only we, but everything – matter, energy, space and time – exists as an integral and interconnected part of a continuum that forms the seamless fabric of the universe, a concept that might be better conveyed by describing the fundamental structure of the universe in terms, not of space–time, but of matter–energy–space–time.

Travellers in time

Even the most serious writers on cosmology seem obliged to consider the issue of time travel, pointing out that it is at least a theoretical possibility. The basis for this is that the laws of physics do not appear to actually prohibit time travel. The agreed mode of time travel seems to be through 'wormholes', sub-microscopic routes out of our universe into another or into another part of the same universe at a different time. The fact that even if one could be found, travellers through wormholes would be turned, as Hawking remarked, into atomic spaghetti seems not to worry these cosmologists overly. However, all seem to agree that time travel as normally understood, if not impossible, is highly impractical. Contrasting travel through time and travel through space, a cosmologist once said that he could travel to the moon but he couldn't walk to next Tuesday. In fact he, along with everyone else, can certainly walk to next Tuesday. They will not be able to do otherwise; we are already time travellers.

As thermodynamic beings in a thermodynamic universe, our very existence is time travel because time is the dimension along which thermodynamic reactions occur. What the cosmologist meant, of course, was that he couldn't travel through time with all other universal thermodynamic reactions continuing as normal, but with the ones that control him, and his ageing, unaltered. He couldn't have his thermodynamic cake and eat it. However, he should be heartened by the fact that just as a trip in a car with someone else deciding the route can be as interesting as one where we decide, so our natural time travel through life, though not within our control, need be no less interesting.

A trip around a circular dimension of time, in which we end up where we started, is the ultimate journey. And, of course, a circular dimension of time makes travelling to *last* Tuesday possible. Unfortunately, just as if we want to get back to the last bus stop, we have to stay on the bus round its circular route once more, to get back to last Tuesday we have to stay on the circular dimension of time. It's a long way round, but we get there in the end. Sadly, we won't be aware of it. After all, we've done it already without noticing it.

The possibility that the dimension of time could be circular changes everything. It means not only the survival of ourselves but also the survival of the human species. It has some speculative spin-offs. The notion that there might be three dimensions of time just as there are three dimensions of space is discussed in Appendix A. The suggestion that the dimension of *scale* may also be circular, so that if you could 'travel' in scale down to the atomic world and keep on travelling, you would end up back at the largest scale, came from string theory, which currently enjoys mixed support amongst theoretical physicists. So these are speculations. But the conclusion of a circular dimension of time with its implications for us is different. It is based on well-established cosmology and scientific principles.

A circular dimension of time might make one think deeply about what to do with the rest of one's life, in order to take full advantage of the opportunity it offers. Or it might be sufficient to know that one can approach the end of one's life with the peace of mind that has been hitherto exclusive to those with total faith in their religion. Either way death is not the end.

Of course, research will continue and may either support or discount this conclusion. That's the way science works. In the meantime, I will happily tell my grandchildren, when they ask what happens when they die, that they will wake up in their mother's arms.

APPENDIX A

A theory of extra dimensions of time

***'The only reason for time is so that everything doesn't happen at once.'* Albert Einstein**

We have seen that quantum theory insisted that a whole range of inaccessible universes could emerge at the Big Bang, one of which is ours, but we were not able to explain how or 'where' these universes could exist, other than in an inaccessible 'meta-universe', which explains little. Three such universes were drawn with a common Big Bang/Big Crunch (Fig A.1). These were not mathematically accurate because, being drawn in the same plane, they would have common points not only at the Big Bang but also at the other intersections. This would mean that there would be accessibility between universes at these intersections, which should not be possible. There are two ways to represent them totally separately from each other. One is with concentric circles in a single plane (Fig A.2), not with a common centre but with a common point on their diameters – the Big Bang/Big Crunch point. The other is by circles at right angles to each other (Fig A.3).

So if we consider all the possible histories but with the same cycle time they would be represented by circles of the same size with the Big Bang/Big Crunch event in common.

If we try to represent this as a physical model, we find that because these cycles are of the same size and have the Big Bang/Big Crunch event in common they can be accommodated within a second dimension of time at right angles to the first and circular (Fig A.4). We can see that at every point on every universe history circle the second dimension is at right angles to the first dimension circle.

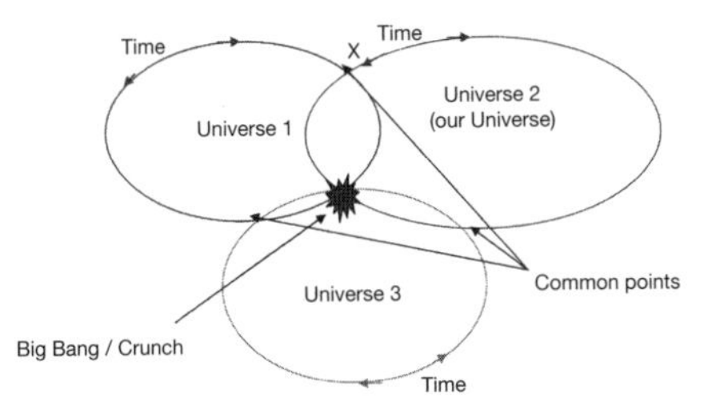

Fig A.1 Three possible universes emerging from the Big Bang, each with a characteristic size and probability of existence. Intersections X, Y and Z are not allowed in a single plane/dimension.

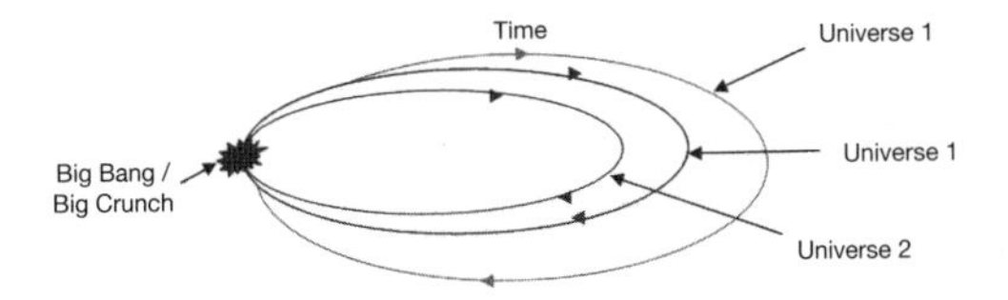

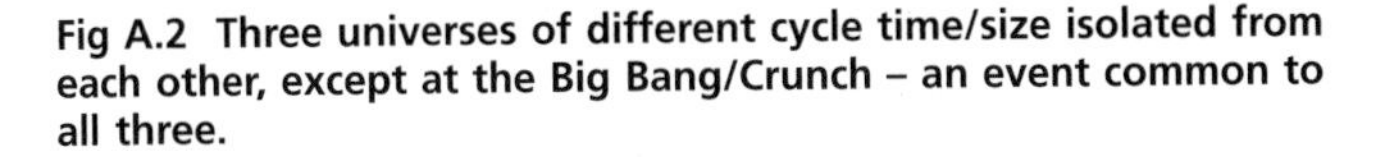

Fig A.2 Three universes of different cycle time/size isolated from each other, except at the Big Bang/Crunch – an event common to all three.

The second dimension, to be at right angles to each time cycle, must also be circular. This creates a time model that is like a doughnut with the Big Bang at the hole that is just closed (Fig A.5). Each cross section around the doughnut represents the circular time dimension of a universe around which its thermodynamic events occur.

These lie in a circle around the doughnut along the second dimension at right angles to each universe dimension. The two dimensions of time are thus represented by the surface of the doughnut. This is reminiscent of Hawking's spherical model whose surface had a similar significance, but which had the Big Bang and Big Crunch as two separate events at the 'north and south poles' of a sphere. In his model, however, Hawking's 'imaginary' time at right angles appeared to have no independent physical significance. Indeed, Hawking seemed to suggest that there *was* no physical significance. It was merely an alternative way of looking at time and a means of making the mathematics work. In this model, the second dimension at right angles to it has the very specific physical significance of accommodating quantum possibilities that otherwise must be conceived of as entirely separate entities, somehow floating about in hyperspace as they were in Hawking's model.

We have seen that there were two different ways of understanding the Big Bang possibilities in quantum theory. We can use Fig A.4 to represent these two different views. The conventional quantum theory view is to see them occurring one after the other as we return to the Big Bang on each cycle. In Fig A.4, this could be represented by the different time dimensions 'flashing' like fairground lights in an unpredictable sequence. In the Many Worlds view, on the other hand, they all exist 'together' alongside one another on a 'super-time' doughnut. These must be understood, however, as equivalent viewpoints, one no more correct than the other.

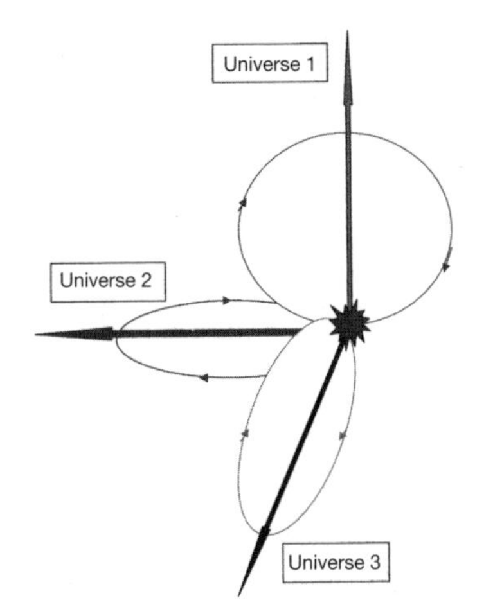

Fig A.3 An alternative way of showing three universes having only the Big Bang/Crunch event in common – in three dimensions at right angles to each other.

A third dimension of time?

Of course, the second dimension of time in Fig A.4 only accommodated the possible quantum universe histories of the same size – all with the same cycle time period as our universe. Although these universes are inaccessible, they are of theoretical interest to us because they are 'twin' to our own – the same size and time cycle period. It is possible that such universes are significant in being the only ones in which life as we know it could evolve. However, smaller or larger universes could emerge from the Big Bang. If these exist, they must not only be independent of the universes like ours but also of each other. They can be accommodated within this model rather neatly.

These different length cycles must obviously be represented by smaller or larger circles (Fig A.2). If we try to fit these circles into our model, we find that we can do so, preserving their independence, if they lie along a third dimension at right angles to the other two (Fig A.6). We thereby create a model that not only accommodates all the quantum history possibilities at right angles to each other, preserving their total isolation from each other as required by quantum theory, but in doing so we have gone further than merely blurring the distinction between time and space; we have created a perfect equivalence between them – three dimensions of space at right angles and three dimensions of time at right angles.

As we go outwards along the third dimension (Fig A.6) we meet larger and larger universes. As we go inwards we meet smaller ones. So although we are considering this as a possible third dimension of time, it must also be a dimension of scale, since it accommodates the different sizes of quantum universe possibilities.

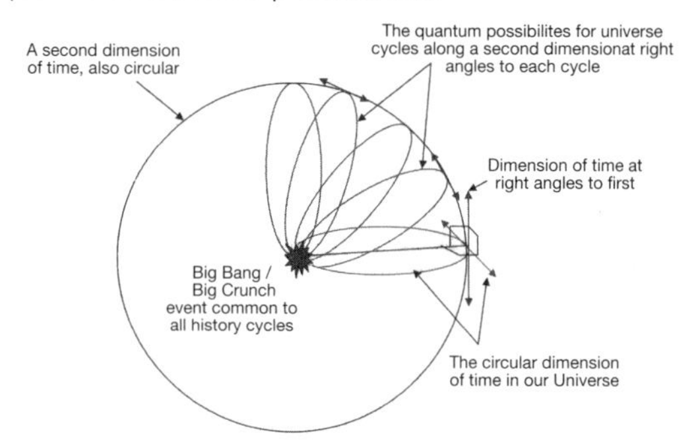

Fig A.4 The quantum possibilities for the universe histories *with the same cycle time*, representing a second dimension of time at right angles to each circular dimension and therefore circular itself.

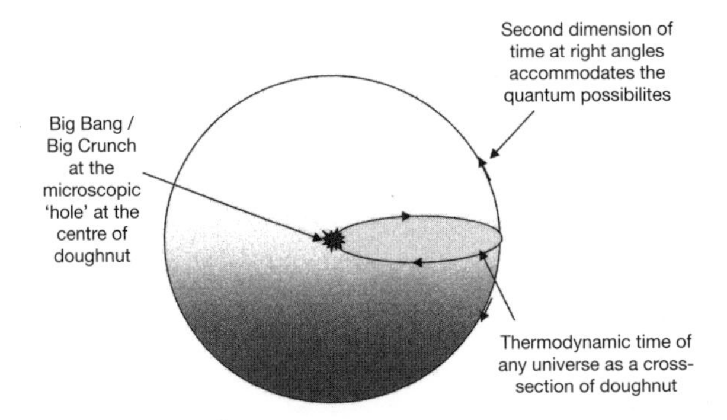

Fig A.5 The two dimensions of time at right angles form the surface of a doughnut. Each cross section around the doughnut represents the circular time dimension of a universe around which its history of thermodynamic events occurs.

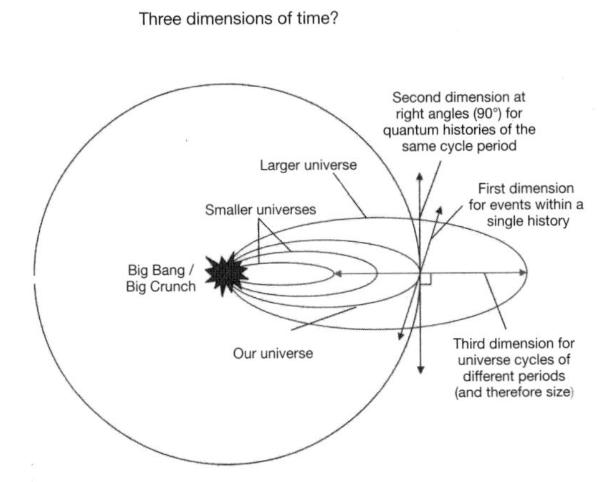

Fig A.6 The quantum possibilities for longer (and therefore larger) universe histories can be accommodated by an extra dimension at right angles to the others like the dimensions of space.

Since the other dimensions are circular in this model, an obvious question arises: could this third dimension, a dimension of scale, also be circular, producing the attractive symmetry of three circular space dimensions and three circular 'time' dimensions? If so, in what way could such a dimension be circular like the other dimensions? This would mean that as we went further out on this dimension, meeting larger and larger universes, at some point we would have to start to come in again, meeting smaller and smaller universes even though we had not changed 'direction' along the scale. Likewise, if we went further inwards along this dimension, the universes would get smaller and smaller and then, at some point, they would somehow have to start to get larger again. This seems absurd, but quantum theory may provide a solution. Quantum theory shows that there is no distance, no point on the dimension of scale, lower than the Planck length. This means that if we could travel down this scale we could get no smaller than this length. As we continued further we would find our path reversed and we would travel back up the scale. So there is certainly a 'loop' at the smallest end of the scale. Could there also be one at the other end, thereby completing the circle? Quantum theory is no help here because it deals only with the smallest scale. At this point, another theory steps in to help us out – string theory. String theory is not so well established as quantum theory, but many physicists argue that it is the most promising attempt to reconcile relativity with quantum theory and discover the 'Theory of Everything'.

String theory proposes that all fundamental particles consist, ultimately, of entities resembling vibrating strings, all made of the same 'stuff', but whose different masses and other properties are determined purely by the nature of their vibration. It proposes, for example, that the universe actually has ten dimensions, of which we have evolved to discern only four: three of space and one of time. String theory appears to be capable of describing the universe as we find it better than any other. One of its strangest findings is that the universe on the smallest scale is physically identical to the universe on the largest scale. Specifically, a closed universe of radius R × the Planck length is physically indistinguishable from a universe of radius 1/R × the Planck length. So string theory, as explained in string theorist Brian Green's book, *The Elegant Universe* (1) shows:

- why quantum theory says that the smallest length that can exist is the Planck length, 10^{-35} metres (incredibly small but not zero),
- why moving down the scale of size, and reaching a minimum, the Planck length, we would seem to be moving back up the scale, even if we had not 'changed direction' and thirdly
- why we could simply not get past the Planck point (the 'R = 1/R equivalence' makes it impossible).

For all quantum universes to be closed, which is the basis of the R = 1/R relationship, they cannot be infinite. Just as they have a minimum size – the Planck length – so they must have a finite maximum size, call it M × the Planck length; huge but not infinite. So, if we imagine scale as the linear dimension that seems so natural to us it would be represented by a line (Fig A.7) with zero size at one end and infinite size at the other. In the diagram, the distance OP – the Planck length – has been exaggerated for clarity. P is actually virtually coincident with O. String theory says that if the maximum size of a closed, collapsing universe is M × the Planck length at point Y, it has an exact physical twin of size 1/M × the Planck length at point X, a universe that is physically indistinguishable. As I argued earlier, if we have two ends of a dimension that are indistinguishable, the Principle of Identity says that they are one and the same point. We can therefore make these points coincident, as we did with the Big Bang and Big Crunch. The result is a circular dimension of scale (Fig A.8). But the reciprocal relationship R = 1/R also means that the length of scale from P to X is indistinguishable from the length of scale from P to Y. So, on the circular scale we could quite properly move P, representing the smallest possible size of universe, to be diametrically opposite Y (Fig A.9). If we now introduce this circular dimension of scale into the model in Fig A.6, the result is shown in Fig A.10, which now has three time dimensions accommodating all the possible universe histories that could emerge from the quantum soup of the Big Bang. They are at right angles to each other, meaning that they are totally inaccessible from each other, except at the Big Bang/Crunch point, and they are circular, mirroring the dimensions of space.

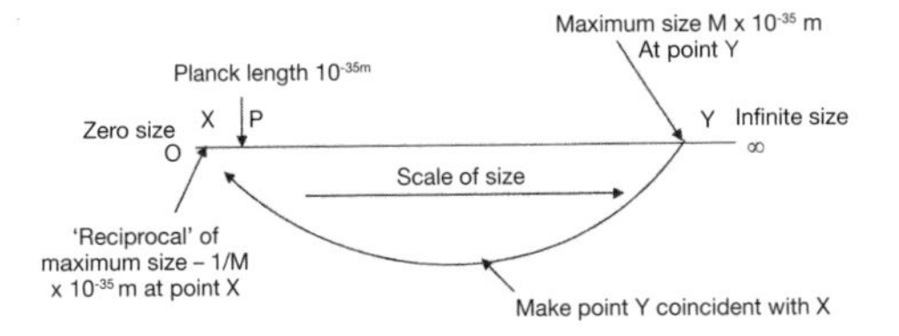

Fig A.7 Scale of size shown as a linear dimension from zero to infinity with the Planck length (exaggerated) close to the zero end. If a universe of size M x 10^{-35} m at Y is physically indistinguishable from a universe of size 1/M x 10^{-35} m at point X, to the left of the Planck point, Y can be brought round as shown to be made coincident with X giving the circular dimension in Fig A.8 (p.160).

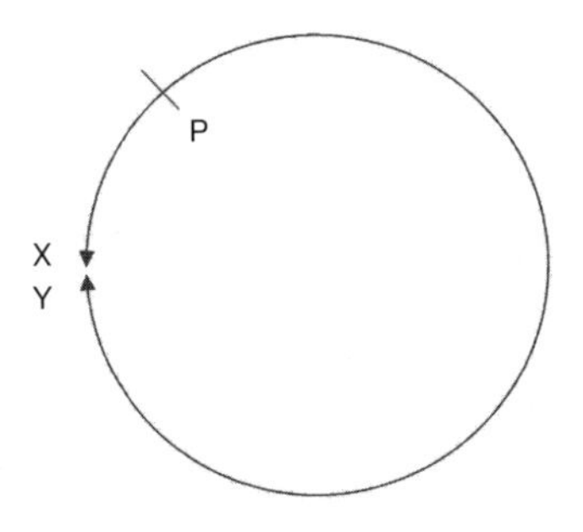

Fig A.8 The circular dimension of scale, which may provide the third dimension of time.

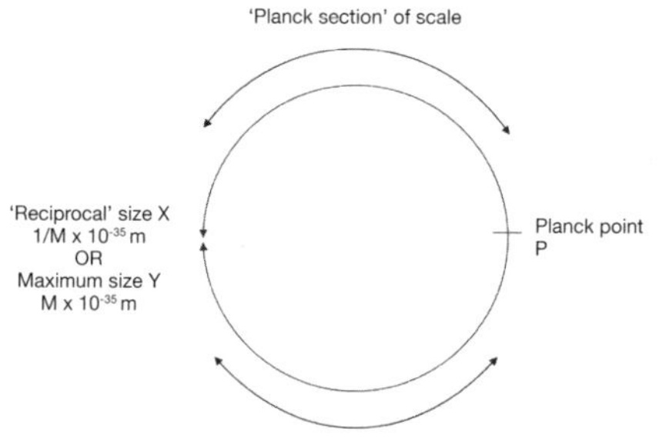

Fig A.9 P moved, showing that the scale from X to P is physically identical to the scale from P to Y as proposed by string theory.

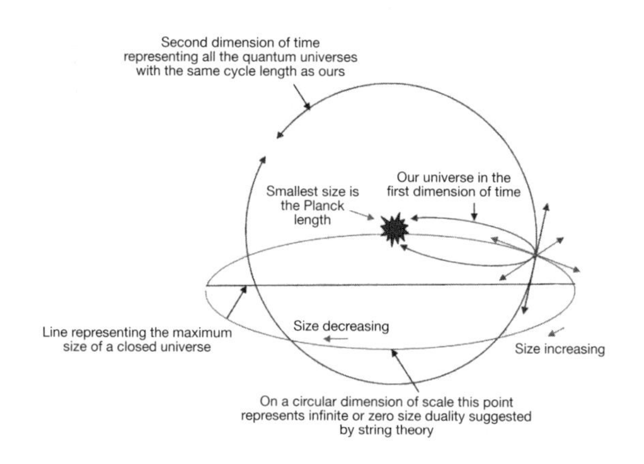

Fig A.10 The third dimension of time accommodating universe cycles with different periods and therefore different sizes represented as a circular dimension.

What could a 'fundamental maximum' size be?

We might ask what the maximum size on the dimension of scale might be. Though there is no physical evidence, it seems reasonable to assume that it is a physical absolute, like the Planck length at the other end of the scale. In view of the reciprocal R = 1/R relationship, we might justifiably speculate that the maximum size possible may be the reciprocal of the smallest size possible (in other words, at the exact opposite end of the scale). Since the Planck length is 10^{-35} metres, this would make the maximum size 10^{35} metres (a 10 followed by 35 zeroes – a huge distance). Recall that cosmic microwave background measurements said that our universe is flat and therefore likely to be one of those with a life long enough to create this flatness. Current measurements of our universe at present suggest that it extends for 10^{26} metres and is expanding. A 10^{35} metre limit – 109 times the current size – would seem to be realistic. It's speculation, but in the absence of any contrary evidence, it appears to be reasonable speculation.

APPENDIX
B

A glimpse into the fascinating quantum world

***'Quantum mechanics: the dreams stuff is made of.'* Steven Wright**

The purpose of science

'Scientific understanding' means that we can predict what will happen because we have worked out the laws that govern what has already happened. Two theories dominate physical science: Einstein's theory of relativity – the theory of gravity and the large-scale structure of the universe – and quantum theory – the theory of the small-scale structure of the universe. The 'Holy Grail' of modern physics is to reconcile these two theories. Relativity theory lends itself most naturally to cosmology, the large-scale behaviour of the universe, and Hawking himself used relativity theory to argue that the universe must have started out with a singularity – a point of zero size and infinite density. Paradoxically and unfortunately, relativity theory is unable to deal with the singularities it predicts, because it cannot deal with infinities, and so it could not offer further insight into the very beginnings of the process. In fact, it has become generally accepted that the normal laws of physics could not apply at such a Big Bang singularity.

The inspired step that Stephen Hawking took was to suggest that the laws of physics did not break down even at this peculiar initial state. He said that the atomic scale of the early universe should mean that not only its initial small-scale behaviour but also its subsequent large-scale development should be governed by the laws of quantum theory. He then set out to determine how.

In doing so he created a model based on the notion of the universe starting off, not from the singularity he had predicted from relativity, but from the simplest state of all – precisely nothing.

The possibility of introducing quantum theory into an area usually described by relativity theory was particularly attractive to Stephen Hawking, as the unification of these two theories into one theory of 'quantum gravity' would be a major step towards his personal mission to find the elusive 'Theory of Everything', and he took bold steps towards it in his model of the universe.

It must be said that some physicists are not happy with Hawking's model. They quibble about his approach and his methods. So whilst Hawking's model does not enjoy universal acceptance, its theoretical pedigree and elegant simplicity convince many that Hawking is on the right track. As American theoretical physicist, Kip Thorne has remarked, 'it smells right' (1).

The strange world of quantum theory

Quantum theory, and particularly its implications, are a scientist's nightmare and a non-scientist's cloud-cuckoo land. It all starts off reasonably enough on the basis that light, well-understood as waves of electromagnetic energy, can also behave like particles – photons. It then gets a bit odd, with the suggestion that matter particles can behave like waves.

Then it gets downright silly, telling us that an atomic particle can be in two places at once and, worst of all for a scientist, that there is a fundamental uncertainty to the physical world that we will never be able to penetrate. At this level, quantum theory limits us to probabilities, not certainties.

A question of size

One of the simplest but strangest implications of quantum theory is that it sets a limit on size. One would imagine that there would be no limit to the smallest size we can conceive of. We might never be able to explore the world at such a scale but it must exist, we might reasonably think. Quantum theory shows that this is not so. If we were able to explore deeper and deeper into the heart of matter at an increasingly smaller scale we would find a fundamental limit at a very small but finite length. There is, according to quantum theory, no length smaller than the so-called 'Planck length'. It's not that we cannot measure anything smaller; it's that nothing smaller exists. The 'Planck length' is 10^{-35} metres (that is, 1/10 with 34 noughts after the 10). There is not only nothing smaller than this; there is no *meaning* to a length smaller than this.

The mystery of the subatomic world

We should not be surprised if quantum theory or relativity seem strange to us. We should expect this because we have not evolved to understand our world at those extreme scales. For example, we understand objects partly in terms of what they look like. But if we consider an electron, the charged particle that forms an essential part of every atom, we have to accept that we can never know what an electron looks like, because it doesn't 'look' like anything. When we say we 'see' an object, we don't see the object, merely a pattern of light reflected whose characteristics we conventionally call the object. But an electron cannot reflect light because it is smaller than the wavelength of light and any other radiation we could use to detect it, which is too 'coarse' to pick out something at the scale of an electron. Even if we tried, the energy in the radiation would so change the behaviour of the electron that we would not be able to see what it was like when we were not looking at it. It would be like trying to see if someone in a darkened room has their eyes open by shining a bright light on them. The light would cause them to close their eyes even if they had been open; the method would be flawed because our intervention changed what we were looking at. So we simply cannot 'picture' an electron; we can only understand it in terms of the things we can detect it doing, and even then we must bear in mind that we are affecting it just by detecting it. So in the end we only know what it does, not what it is. But then, of course, if we only know of its existence when it is doing something, then we cannot know anything about it when it isn't doing anything. It's rather like seeing a famous public figure on television. We see only their 'image'. We cannot see on television what they are like when they are not on television. However, whereas we could devise ways of finding this out, with subatomic particles there is no way of knowing anything about them when we are not detecting them doing something, nor, therefore, for predicting precisely, after they have done something, what exactly they will do next. This has important consequences for the way in which we have to deal with them and shows why quantum theory must deal with probability not certainty.

To get some sort of understanding of this extreme level of scale, we must resort to mathematical – rather than physical – models to give us the most accurate description. The model of an atom we learnt at school with 'planetary' electrons spinning around a nucleus is still useful for a simple understanding of chemical reactions or conductivity, but that's about all. This is why Hawking, like other cosmologists, is much more likely to be describing an atom – or the whole universe – with a page full of equations than with a picture.

One advantage of Hawking's model of the universe is that, although it was developed mathematically, it can be represented visually. And, although it is based on quantum theory, it does not require a deep understanding of quantum theory to appreciate important aspects of it but merely a grasp of certain principles that are fascinating rather than daunting, especially as mathematics is not needed.

Atomic particles or atomic waves?

The first thing to keep in mind is that, since quantum theory describes the working of our physical world at the level of elementary particles, by the time we have reached that level, our common sense view of the world has long broken down. We must therefore not expect things to be in any way 'normal'.

For example, looking again at the electron, we normally think of it as an electrically charged particle. We use that word because electrons behave like particles. They move about like particles and we can 'track' them. Jumping from atom to atom, they are responsible for chemical reactions.

In conductors they also jump from atom to atom, but in a constant stream as an electric current. And, like any other particle, we can weigh an electron. It weighs about 1/2000 of a hydrogen atom. On the other hand, electrons can also behave like something not remotely like a particle. If beams of electrons are shone through two adjacent slits, they behave like light waves. When light is passed through two slits close together, the beam splits into two new beams that interfere with each other to produce a pattern of dark and light bands on a screen – an 'interference' pattern. It is an exclusive feature of waves. English physicist Thomas Young used this effect in 1801 to establish the wave nature of light in the face of the prevalent view at the time that light was made up of particles. The same effect is seen in water waves, which pass between two gaps close together. Now we see that electrons do exactly the same thing.

So are electrons particles or are they waves? In fact they are neither – or both. As we saw above, we cannot say what electrons look like, only what they do, and then only when we are watching them. But now we see that what they do depends on how we look at them. The interference experiment shows that electrons behave like particles when we look at them one way, and like waves when we look at them in another way. How they behave at any time depends on what they are asked to do. In quantum physics this is called 'wave–particle' duality. This is not so strange as it sounds. A familiar object, when viewed from different perspectives – say, at right angles (90°) to each other – can look very different. A pyramid looks like a triangle from the side and a square from below. A cylinder looks like a circle from the end but a rectangle from the side. But we don't ask whether a pyramid is triangular or square, or whether a cylinder is circular or rectangular. This is because we are able to see things from every viewpoint. This is normally the case with everything we look at. Unfortunately, with an elementary particle like an electron we have only two viewpoints and they are mutually exclusive. One viewpoint sees particles, the other sees waves. Neither 'particle' nor 'wave' is a totally adequate description but there are only those two ways of looking at them. Like the pyramid, which is more than a triangle or a square, electrons may actually be something 'more' than particles or waves, but since we don't have anything but particles and waves to compare them with we do not know. So we simply talk about 'wave–particle duality'. Just as we cannot see a cylinder from the top and the side simultaneously, we can never see electrons behaving like particles and waves at the same time, so we can only treat them according to the way they are behaving when we are dealing with them.

Amazingly, the wave properties of matter were predicted in 1924 by French physicist Louis de Broglie (pronounced 'de Broy') – before they were actually discovered. Einstein had earlier shown that light, then understood to be simply continuous waves of electromagnetic energy, was made up of streams of small 'particles' of energy – photons. De Broglie boldly proposed the same dual nature for *all* matter, and backed it up with elegant mathematics. Just as light must be considered to have both wave and particle properties, so matter, said de Broglie, must be regarded as having both particle and wave properties. It was a brilliant *tour de force*. He was initially faced with scepticism from the scientific community but this evaporated when electron waves were discovered. Einstein and de Broglie duly received the Nobel Prize for their achievements.

De Broglie showed that all matter has wave properties. In due course, wave properties were discovered in neutrons – the nuclear particles with two thousand times the mass of electrons – and subsequently in whole atoms. De Broglie showed, however, that the larger the lump of matter we look at, the smaller and less significant are its waves. This is the reason that we aren't being tossed around on an ocean of matter waves. It's a bit like waves on the sea. On a small scale, the waves are very significant (think of sitting in a dinghy in the Atlantic). On a larger scale the individual waves, however rough the sea, have little effect on the sea as a whole. From a satellite, the turbulent sea around Cape Horn looks very much the same as the calm sea of the Mediterranean. Indeed, from the satellite the waves might just as well not exist. Individually they have little or no relevance to the behaviour of the oceans as a whole. In the same way, the waves of electrons are very significant in understanding their individual behaviour. For atoms they are less so, and for matter on an everyday 'macro' level the waves are not significant at all. The wave properties of a star, for example, are so infinitesimally small that they can be ignored. At an atomic level they are so large that they must not be ignored.

The wave nature of matter introduces a fundamental uncertainty into our powers of prediction. A wave stretches on forever; it has no 'position'. The wave of an electron represents an uncertainty about where it is and what it is doing. It means that when we try to pin down the position of an electron exactly, we lose our ability to say what it was doing when we found it and what it's going to do next. At the subatomic level, where the wave properties are at their most significant, it is impossible, even in principle, to predict exactly. We can only predict in terms of probability. The waves of matter in fact represent waves of probability. This essential uncertainty is expressed by a fundamental law of quantum physics derived by German physicist Werner Heisenberg and his colleagues, known as the 'uncertainty principle'. It says that it is absolutely impossible to determine the exact state of anything at the quantum level, like an electron. If you know its position exactly, you cannot know how it is moving. If you know how it is moving, you cannot know where it is. This has fundamental implications. For example, it means that no particle, whether an electron or an atom, can be stationary, for then we would know its position as well as its motion. This explains why, for example, it is impossible to reach a temperature of absolute zero at which particles would have zero energy and therefore be stationary. German physicist Hermann Nernst had earlier proved the same thing using a thermodynamic approach, a good example of the confirmation of quantum theory.

Since large-scale matter has insignificant waves, the limitations set by the uncertainty principle at this level are also insignificant. Large-scale theories therefore predict exactly. So astronomers can tell us exactly when Halley's comet will return and chemists can tell us exactly how much sodium chloride will be produced when sodium and chlorine gas react together.

On the quantum scale at which the universe began, however, the very essence of science, the prediction of physical processes can only be made in terms of possibilities and probabilities. For example, suppose an electron is detected at point A. Quantum theory says that it is absolutely impossible to predict its subsequent behaviour exactly. And even if it arrives at point B we cannot know the route it has taken, only the probability of each possible path. The path taken is sometimes referred to as the electron's 'history'.

A helpful way of getting a feel for the quantum view of uncertainty is to think of the behaviour of ants. Although there is no wave behaviour – the ants can only behave like particles – they are always moving, like electrons. If we saw an ant at some arbitrary point A within its habitat and subsequently at point B, and we did not see it at any other time, we would not, could not, know the behaviour, the 'history', of the ant before it got to A, in going from A to B, or after it leaves B. All we could do is to plot the paths of all the ants we could see over a period of time and work out probabilities about their behaviour. If at one time we saw an ant in one place, when we looked again later we would expect to find it somewhere near, but in fact it could be anywhere. There would be a smaller probability of finding it a long way from A but so long as it was possible, the ant could be absolutely anywhere. If we waited long enough, our ant would take every possible path. This corresponds exactly to the behaviour of an electron or any quantum particle. Every possibility, anything not actually forbidden, will occur. This is an important principle, which we will return to later.

We could make the ant behave even more like an electron by applying a 'field' – a pulling force – between

A and B – which has the same effect as an electric field on electrons. If electrons are placed between two metal plates to which a voltage is applied, the electrons drift in the direction of the resulting electric field. If we make B a supply of sugar, and A the entrance to the ants' nest, we create an 'antic' field and there will be an initial general drift from A to B, just as electrons would drift in the electric field. The ants would take different paths, and the probabilities of different paths would vary, but the highest probability path would be a straight line between A and B. Notice that whilst the probabilities would change, the possibilities would remain the same. Notice, too, that although the behaviour of the individual ants would be uncertain, the behaviour of the ants as a whole would be very predictable. A video of ant movement taken at one point in time would be virtually indistinguishable from a video taken at any other time.

From any initial state, then, we must think of a quantum particle, like an electron, following every path allowed, each with a certain probability, and not simply confined to one, albeit unknown, path. This is not confined to single particles, however. Any quantum system, any group of subatomic particles, must be thought of as having every possible subsequent state, each with a certain probability. So although the system is governed by uncertainty, the probabilities themselves are certain. All we need to know is how to work out these probabilities.

The good news is that the very wave properties of particles that cause all the problems in fact allow a specific mathematical approach to be applied to measure the probabilities we need to know. This approach, developed by Austrian physicist Erwin Schrödinger in the early 1930s and known as 'wave mechanics', is a fundamental part of quantum theory. Here, the behaviour of a quantum particle is described mathematically by its 'wave function', which measures the probabilities of its behaviour. The wave function of an electron in a hydrogen atom, for example, describes it as a probability cloud around the nucleus, not as a particle in a spherical orbit around the nucleus like a planet around the sun, as once believed in 'classical' physics. The greatest probability density, however, coincides with the classical 'planetary' orbit. In fact, wave mechanics represents the path of any quantum particle by a cloud of probability quantified by the electron's wave function.

The problem comes about in explaining what happens when, say, an individual electron actually turns up. Until then we have been obliged to consider it in terms of a probability cloud described by its wave function. What happens to the probability cloud and the wave function when we finally get the electron in our sights? The conventional view of quantum theorists, established in the late 1920s by Niels Bohr and colleagues at the University of Copenhagen, is that the effect of our observation makes one possibility 'crystallize out' of the probability cloud in much the same way as, when a photographer asks a group of people to look his way, they stop all their different behaviours and adopt just one, grinning at the camera. Mathematically, this is called the 'collapse of the wave function'. Unfortunately, this introduces the idea that an electron, like an ant, can, with an appropriate probability, be anywhere, doing anything, until we look at it. This is not a notion that scientists naturally feel comfortable with.

However, this conventional way of understanding quantum behaviour, known as the 'Copenhagen convention', has been almost universally accepted because it works perfectly well. It's actually not quite so strange an idea as it first seems. A lottery draw works this way. Until the final selection, all the numbers exist as equally probable possibilities. The winning number is obviously there, but there is no way of knowing what it is until we look at it. Only then does it effectively 'crystallize' out of all the possibilities into the reality of the winning number.

An alternative view of the quantum possibilities, avoiding the need to have wave functions collapsing and probabilities crystallizing out, is that all the possibilities somehow occur together as parallel universes. This idea represents an interpretation of quantum physics called the 'Many Worlds Interpretation', which is entirely consistent with the conventional view and is accepted as a perfectly legitimate, if not widely held, interpretation. It was presented in heavy mathematical form in 1957 by American physicist Hugh Everett. It was largely ignored at the time, even though it was given its seal of legitimacy by one of the most prominent quantum physicists, the late John Wheeler. It is unfortunate that Everett did not live to see his theory taken seriously by quantum cosmologists, for it solved a problem not addressed by Niels Bohr's Copenhagen interpretation. In Everett's view, all the quantum possibilities for a particle occur by the universe splitting into different 'worlds', in each of which every quantum possibility becomes the reality in that world. Each world thereafter begins a separate history, unconnected to every other. Mathematically, we could say that these worlds therefore go off 'at right angles' to each other. Stephen Hawking has a soft spot for this interpretation of quantum physics, incorporating it into his no-boundary model.

Wave mechanics do not only apply to single particles, however. Any quantum system, any system of atomic scale particles, can be described mathematically by a wave function, to give information about the system and its future behaviour – its 'history' – but only, of course, in terms of possibilities and probabilities. Stephen Hawking made the simple but inspired observation that since the whole of the universe started out, according to accepted 'Big Bang' theory, at a sub-atomic scale, its subsequent behaviour should be understandable in terms of quantum laws. Like any other quantum system, this means that at its initial 'Big Bang' quantum state, the universe had not one, but a whole range of possible histories open to it. Each had a probability determined by the wave function of the initial quantum-state universe. We know for sure, of course, what one of these possible histories was – we are now living in it. But just because it is the one that we know actually happened doesn't mean it was a high probability one. Quantum theory says that it could be a very low probability one. Without an understanding of the wave function of the initial universe, we cannot know.

Stephen Hawking was bold enough to attempt to determine the wave function of the entire universe at the Big Bang. He took each of the cosmological models predicted by Einstein's relativity theory and, with James Hartle, assigned a wave function to them, thereby creating a set of mathematical models of the universe.

The beauty of Hawking's mathematical wave function model, as we have seen, is that all these wave models could all be represented by a single, physical, visual model. Hawking did not make the one small extra step of seeing the Big Bang and Big Crunch as a single point. We might ask why he didn't do so. The obvious answer is that he did consider it and rejected it. But had he done so he would surely have told us about it, and he didn't. Perhaps he was so consumed by the minutiae and complexity of his ground-breaking theory that he overlooked this simple possibility.

REFERENCES

Introduction

1. Paul Davies in Henblest, Couper, *Extreme Universe,* Channel 4 Books, (2001), p 174
2. Stephen W. Hawking, *A Brief History of Time*, Bantam, London, (1988), p 187
3. Ditto, p 154

Chapter 1 Our universe – what it's like and how it got to where it is today

1. Simon Driver, Quoted in *The Daily Telegraph*, 24 July 2003

Chapter 2 The origin of the universe – a universe from nothing?

1. Edward Tryon, '*Is the Universe a Vacuum Fluctuation*?', Nature, Vol 246, (1973)
2. Alan H. Guth, *The Inflationary Universe,* Vintage, London (1998)
3. A. Vilenkin, '*Creation of Universes from Nothing*', Physics Letters, Vol 117B (1982)
4. Stephen W. Hawking, *Black Holes and Baby Universes,* Bantam, London, (1993), p 97
5. Paul Davies, *The Last Three Minutes,* Phoenix, London, (1995), p 167

Chapter 3 Hawking, his universe and the first hint of circularity

1. Stephen W. Hawking, *A Brief History of Time*, Bantam, London, (1988), p 155
2. Stephen W. Hawking, *Black Holes and Baby Universes,* Bantam, London, (1993), p 36
3. Stephen W. Hawking, *A Brief History of Time*, Bantam, London, (1988), p 153
4. Stephen W. Hawking, *A Brief History of Time*, Bantam, London, (1988), p 154

Chapter 4 From imaginary time to circular time – the Principle of Identity

1. Stephen W. Hawking, *A Brief History of Time*, Bantam, London, (1988), p 193

Chapter 6 Storm clouds brewing for the closed universe and the Big Crunch

1. Paul Davies, *The Last Three Minutes,* Phoenix, London, (1995), p 135
2. Robert Dicke quoted in Alan H. Guth, *The Inflationary Universe*, Vintage, London (1998), p 19

Chapter 7 Supernovae send a game-changing message from the past

1. Paul Davies, The *Last Three Minutes,* Phoenix, London, (1995), p 113

2. Paul Davies in Henblest, Couper, *Extreme Universe,* Channel 4 Books, (2001), p 174
3. Max Tegmark, Scientific American, (May 2003)
4. Paul Davies, Personal Communication (Nov 24, 2005)
5. Don Page, *Information Loss in Black Holes and/or conscious beings,* Institute for Theoretical Physics, University of Alberta, (Nov 25, 1994)

Chapter 8 What exactly makes you *you*? – the meaning of 'consciousness'.

1. Herman Minski, *The Society of Mind,* Heinemann, (1987)
2. Derek Parfitt, *Reasons and Purpose,* Oxford University Press, Oxford, (1984)
3. Herman Minski, *The Society of Mind*, Heinemann, (1987)

Chapter 9 Holes appear in the accelerating universe theory

1. Paul Steinhardt, Neil Turok, *Endless Universe*, Weidenfeld & Nicolson, (2007)
2. Roger Penrose, *Cycles of Time*, Vintage, (2011),
3. Nemanja Kaloper and Antonio Padilla, http://phys.org/news /2015-03-universe-brink-collapse-cosmological-timescale.html
4. Peter Milne, see http://www.eurekalert.org/pubreleases/ 2015-04/uoa-aun041015.php,
5. Christos Tsagas, *Peculiar motions, accelerated expansion and the cosmological axis* Physics Review, D84, 063503, Sep 2011
6. Christos Tsagas, ditto
7. Janna Levin, *How the Universe got its Spots*, Phoenix, London, 2002
8. Lawrence Krauss, *A Universe from Nothing*, Simon & Schuster, London (2012) p147

Chapter 11 Science and simplicity

1. Stephen W. Hawking, *Black Holes and Baby Universes,* Bantam, London, (1993), p 155
2. Alan H. Guth, *The Inflationary Universe*, Vintage, London (1998), p 238
3. John Gribbin, *In Search of the Big Bang,* Penguin, London, (1998), p 331
4. Richard Panek, *The 4% Universe*, One World, Oxford,(2011)
5. Stephen W. Hawking, *A Brief History of Time*, Bantam, London, (1988), p 193

Chapter 12 Summary and conclusion: the implications of a circular dimension of time

1. Stephen W. Hawking, *Black Holes and Baby Universes*, Bantam, London, (1993), p 139
2. Stephen W. Hawking, *Black Holes and Baby Universes,* Bantam, London, (1993), p 135
3. Isaac Asimov, *Asimov's New Guide to Science,* Penguin, London, (1987), p 378
4. Eric Rogers, *Physics for the Enquiring Mind,* OEP, Oxford, (1960) (Out of print) p 751
5. John Muir, quoted by John Nichols in Natural History, November 1992

Appendix A

6. Brian Greene, The *Elegant Universe*, Vintage, London, (2000) p 248

BIBLIOGRAPHY

This book could not have been written without access to the work of a large number of authors, whose knowledge and ability to communicate this knowledge have been inspiring. Any parts of this book which subsequent research shows to be wrong are my responsibility not theirs. For the most part these books are not only accessible, with little or no mathematics, but they are entertaining as well as informative, giving us insight into some of the most intelligent and interesting people on our planet. These authors are:

Stephen W. Hawking, ***A Brief History of Time,*** Bantam, London, (1988)

The book in which Stephen Hawking first made his work on cosmology accessible to the public.

Stephen W. Hawking, **Black Holes and Baby Universes,** Bantam, London, (1993)

A collection of transcripts of Stephen Hawking's lectures and articles from 1977 to 1992 ending with an interview for the *Desert Island Discs* radio programme on Christmas Day 1992. Hawking gives some of the background to his work and expands on the ideas presented in *A Brief History*. He tries to clarify 'imaginary time' and discusses determinism on a scientific basis.

Stephen Hawking (Ed), ***A Brief History of Time*: A Readers Companion,** Bantam, London, (1992)

This is written as a companion to the documentary film about Hawking and his work, with personal contributions from many notable cosmologists, some of whom were his contemporaries at Cambridge. It offers insights into Hawking's background, his family, his early work, and the way his work is viewed by others. Again, some of the more difficult parts of *A Brief History* are further illuminated.

Stephen Hawking, **The Universe in a Nutshell,** Bantam, London, (2001)

This is Stephen Hawking's follow-up to *A Brief History* with a wealth of fine illustrations. He goes over some of the concepts in *A Brief History* and expands others. Imaginary time remains obscure, as do the universe models constructed from it.

David Wilkinson, **God, the Big Bang, and Stephen Hawking,** Monarch, Crowborough, (1996)

This could have been called *A Brief History of Time* – a Religious Perspective. The author, a prominent astrophysicist, presents much of the science of *A Brief History* clearly and concisely with further illuminating background information.

Alan H. Guth, **The Inflationary Universe,** Vintage, London (1998)

Guth was responsible for the theory of rapid inflation of the universe to solve some of the most intractable problems presented by the Big Bang theory – 'the ultimate free lunch'. In this book he charts in detail the path towards his theory, reviewing the background and discussing the issues, providing a fascinating insight into key research work in cosmology, past and present. He explains how the positive matter–energy in the universe can be cancelled by negative gravitational energy.

Brian Greene, **The Elegant Universe,** Vintage, London, (2000)

An account of the basis of superstring theory, its contribution to current cosmology and the basis for the belief that it holds the key to a complete understanding of the universe – the elusive Theory of Everything.

John Gribbin, **In Search of the Big Bang,** Penguin, London, (1998)

This book, in its first, 1986, edition, made the greatest impact on me after reading *A Brief History*. The author's scientific knowledge is only matched by his skills as a writer and communicator. Apart from the solid science, what comes over is his empathy with the lay reader and his understanding of the processes and problems of learning science.

Isaac Asimov, **Asimov's New Guide to Science,** Penguin, London, (1987)

The definitive and comprehensive lay person's guide to science. Clearly and interestingly written, it gives wonderful insights and a wealth of information.

Paul Davies, **The Last Three Minutes,** Phoenix, London, (1995)

This has a slightly different perspective, charting and explaining the possible scenarios of the future of the universe.

J.P. McEvoy and Oscar Zarate, **Stephen Hawking for Beginners**, Icon, Cambridge, (1995)

A concise explanation of Hawking's background and the physics behind his work. Heavily illustrated, it is the only book that ventures to illustrate Hawking's 'real' and 'imaginary' time as components of time at right angles to one another.

Nigel Henbest & Heather Couper, **Extreme Universe,** Channel 4 Books, London, (2002)

This is the accompaniment to the three-part Channel 4 series of the same name. It is an easy-reading, broad look at the universe and includes conversations with a number of cosmologists discussing some of the latest research. Of particular interest is the final chapter in which Janna Levin predicts a closed, finite universe in the face of current cosmological opinion (see also her book, below) and Paul Davies suggests that an infinite, open universe may still offer the possibility of a repeat of our existence. Interestingly he considers a universe infinite in space, in which we must be repeated. He does not also consider a universe in time, which must have the same result.

Janna Levin, **How the Universe Got its Spots,** Weidenfeld & Nicholson, London, (2002)

The author blends serious cosmology with a candid diary based on letters to her mother over a two-year period. Despite the recent evidence that convinces most cosmologists that the universe is infinite, Ms Levin argues that the universe is in fact finite, and draws conclusions which add weight to the possibilities that are set out in this book.

Richard Panek, **The 4% Universe,** One World, Oxford,(2011)

This tells the story of cosmology from 1965 to the present day in amazing detail. It's as if the author sat in on every discussion by every cosmologist. A remarkable book.

Lawrence Krauss, **A Universe from Nothing,** Simon & Schuster, London (2012)

The author's enthusiasm, seen in his many videos, shines through the pages of this book. His description and explanation of a universe from nothing made a significant contribution to this book

John D. Barrow, **The Infinite book,** Vintage, London, 2005

The author says most of what anyone needs to know about infinity and its implications.

ACKNOWLEDGEMENTS

I should like to thank the following people for their huge contributions to the completion of this book: Tessa Rose, Niko Heyng and Daniel Conway at Arcturus for their interest and faith in this book, and for their patience, support and understanding through its preparation; Richard Emerson for his critical eye, pertinent observations and advice; all those authors who, with enthusiasm characteristic of their profession, have given me the understanding of cosmology that I lacked before starting this book, in which any errors are mine alone; Anna van Lidth for her time, advice and invaluable practical help, without which this book would have been much the poorer; and, for their interest and encouragement, my friends and family, especially my wife to whom I dedicate this book.

INDEX